What People A

Infinite Perception

Humankind is now facing major existential threats. The emergence of global consciousness is absolutely required to solve these threats. If you want to learn more about how psychedelics can foster such emergence and positively contribute to individual, societal, and cultural transformations, please read this important and timely book.

Mario Beauregard, neuroscientist, author of *Brain Wars* and *Expanding Reality*, co-author of the "Manifesto for a Post-Materialist Science"

Infinite Perception is a brilliant new anthology of timeless wisdom for these changing times. As the weaponization of culture filters our consciousness, just as psychedelics penetrate the capitalist mainstream, we need voices—and experience—like the authors here share, that understand the vision emerging from our altered states is a vital part of the global transformation. Beyond the medical model of psychedelics, the mystical experience is what connects us to meaning and understanding our place. Indigenous voices parallel Western doctors and explorers in this vital collection that maps the cultural and societal transformation afoot with psychedelics, and the personal transformations that heal and illuminate our journeys into the mystery.

Rak Razam, author of *Aya Awakenings*, filmmaker

Infinite Perception

The Role of Psychedelics in
Global Transformation

Previous Books

Expanding Science: Visions of a Postmaterialist Paradigm
Edited by Mario Beauregard, PhD; Gary E. Schwartz, PhD;
Natalie L. Dyer, PhD; and Marjorie Woollacott, PhD
ISBN-10: 1-73544-912-1
ISBN-13: 978-1-73544-912-8

Moon Cartagena and Colombia's Caribbean Coast
By Ocean Malandra
Publisher Moon Travel
ISBN-13: 978-1-64049-941-6

Infinite Perception

The Role of Psychedelics in Global Transformation

Edited by

Ocean Malandra

and

Natalie L. Dyer, PhD

BOOKS

London, UK
Washington, DC, USA

CollectiveInk

First published by O-Books, 2024
O-Books is an imprint of Collective Ink Ltd.,
Unit 11, Shepperton House, 89 Shepperton Road, London, N1 3DF
office@collectiveinkbooks.com
www.collectiveinkbooks.com
www.o-books.com

For distributor details and how to order please visit the 'Ordering' section on our website.

Text copyright: Ocean Malandra and Natalie L. Dyer, PhD, 2023

ISBN: 978 1 80341 460 7
978 1 80341 461 4 (ebook)
Library of Congress Control Number: 9781803414607

A CIP catalogue record for this book is available from the British Library.

Design: Lapiz Digital Services

UK: Printed and bound by CPI Group (UK) Ltd, Croydon, CR0 4YY
Printed in North America by CPI GPS partners

The authors of this book do not dispense medical advice or prescribe the use of any technique as a form of treatment for physical, emotional, or medical problems without the advice of a physician, either directly or indirectly. The intent of the authors is only to offer information of a general nature to help you in your quest for emotional and spiritual well-being. In the event you use any of the information in this book for yourself, which is your constitutional right, the authors and the publisher assume no responsibility for your actions.

We operate a distinctive and ethical publishing philosophy in all areas of our business, from our global network of authors to production and worldwide distribution.

Contents

Ocean Malandra is a freelance writer from Northern California that has spent much of the last two decades traveling and living in Latin America, much of that time in the Amazonian regions. His work focuses on solution-based environmental journalism, sustainable food and travel, social justice, and psychedelic plant medicines. He has been published in over 30 different media outlets including Mongabay, VICE, *Earth Island Journal*, *Parabola* magazine, *High Times* magazine, *Amazon Watch*, Leafly, Reset.me, Invisible People, and *Paste* magazine, where he wrote the environmental column EarthRx. Ocean is also the author of the Moon Travel guide series on Colombia, a country that has become his home away from home.

Contact: oceanmalandra@gmail.com

Dr. Natalie L. Dyer, PhD, is a Research Scientist with Connor Whole Health at University Hospitals, President of the Center for Reiki Research, and serves on the board of the Scientific and Medical Network. Natalie completed her doctorate in neuroscience at Queen's University and postdoctoral fellowships in psychology at Harvard University and Harvard Medical School.

She has published many scientific papers and book chapters on postmaterialist science, psychedelics, and the therapeutic effects of integrative medicine practices, including yoga, acupuncture, mindfulness, and energy medicine. She is co-editor of the book *Expanding Science: Visions of a Postmaterialist Paradigm*. Natalie is also an energy medicine practitioner and teacher with clients around the world. Her passion is in understanding and addressing mental and physical health from a psychospiritual perspective. She lives in the woods on the east coast of Canada with her husband, artist Louis Dyer, and their daughter.

Website: drnataliedyer.com

Contact: natalie.leigh.dyer@gmail.com

Introduction

Into the Mystic

Ocean Malandra

We are now fully in the middle of a "psychedelic renaissance." After decades of being pushed into the underground, visionary plants and other mind-opening substances are not only being studied at major universities all around the world, but at the grassroots level, communities, cities, and even entire states are decriminalizing their use. Behind this global movement lies a belief that psychedelics have a unique role to play in transforming the world by facilitating change right at the most primary level—human consciousness.

How that will happen and what will result from it is the theme of this book.

We believe that psychedelics are indeed a powerful tool for helping to facilitate transformation at every level of our now global civilization. The root meaning of "Psyche," which is "mind" or "soul" plus the root meaning of "Delic," to "manifest," literally points to their potential in helping us dream up new possibilities and make them realities. That is what this book is about.

The mainstreaming of psychedelics has also attracted certain interests, motivated by the investment potential, who are attempting to dominate the conversation with promises of "psychedelic therapies" to treat modern crisis level issues like depression and addiction. This book will also widen the conversation around that deceptively simple narrative.

While psychedelic compounds can be shown to have biochemical effects on the brain, what dozens of studies actually show[1] (and tons of personal stories also attest to) is that it is the extent that each person feels like they have had a "mystical experience" while under the influence of psychedelics that determines whether they get a therapeutic effect.

In other words, the magic happens at the consciousness level, not just the chemical. True to their literal definition, psychedelics are conscious change agents, not simply pharmaceuticals with only mechanical modes of action. In other words, they are active medicines, not passive. They need your participation to be effective. This is the core reason why all the hyped-up pharmaceutical claims fall flat. A theme that will be explored extensively in this book.

After all, studies are now showing that most of our mental health epidemic is being driven by social forces and not chemical ones. Depression and addiction shoot up when student and credit loan debts climb,[2] when soulless jobs and lack of community leave people feeling isolated, and when the stress of capitalistic competition in every arena of life leaves one feeling drained and numb.

In fact, a recent study proved that when depression is not treated as an abnormal response to a normal situation, and instead seen as the opposite—a normal response to an abnormal situation—patients immediately felt better and their outcomes actually improved.[3] This turns the Western model of mental health on its head, literally. And it's about time.

With scientists telling us that the ecosystems of our planet simply cannot sustain any more business as usual, with global superpowers poised on the edge of World War III, and unprecedented inequality stretching social tensions to their breaking point all around the world, I think we are finally ready to collectively face reality. To me, this is the real reason why the "psychedelic renaissance" is happening right now.

Psychedelics are not about simply numbing the symptoms so we can get back to work on a system that we should be moving beyond. They are for opening the mind to other possibilities.

They are for entering that mystical realm, where true wisdom is gained that can truly transform our world.

This is what has inspired the title of this book. It comes from a well-known 19th century poem by William Blake, which states that:

If the doors of perception were cleansed, everything would appear as it really is, infinite.

While this phrase has become a kind of recurring reference point in modern psychedelic culture—everything from Aldous Huxley's *The Doors of Perception*, which details his explorations with mescaline, to the epitome of countercultural 60s musical band The Doors—Blake himself was actually inspired to write it by Plato's "Parable of The Cave," as Professor Carl Ruck explains in chapter 7 of this book.

Written at the dawn of Western civilization, the parable of the cave is a starkly powerful description of the limited perception we experience in everyday life. A reality in which the intricate connections between things, the infinite connections even, are not always apparent. Trapped in the darkness of a cave, stumbling in the near blindness of our own limited awareness, we don't have the ability to see exactly how we have gotten ourselves into our dilemmas, both personal and global, and we have a hard time seeing what "other possibilities" exist.

In Plato's allegory, one of the cave dwellers makes their way out into the sunshine. A reality of only vagueness and shadows is instantly transformed, illuminated, into a world of full spectrum colors, arching skies and endless horizons, populated with all the various forms of life that the world contains. It is no wonder that this allegory, and Blake's poeticized version of it, have become metaphors for the psychedelic experience.

Under the influence of psychedelic plant medicines, the unseen becomes vividly real, often even more real than reality itself as it pulses in technicolor visions before us. It is as if in the middle of the night, lost in the dark forest, we are suddenly given the night-vision of an owl and the whole world in all its intricately detailed nuances, only hinted at in the shadows before, is now cast in the floodlights for us to inspect, up close, and verify with our own senses.

This is the mystical experience that brings real healing. Here in the illuminated reality of infinite perception, where everything is interconnected, interdependent, and intertwined, every one of our actions is shown to have a ripple effect, not just among those that they immediately touch, but across the entire cosmos. It's like turning a tapestry over and seeing that, from the back, it's apparent that all those different figures and shapes are really all created from the same thread. A thread that weaves its way in and out of visible reality in a constant continuum.

A thread that has no beginning and no end. This is the mystical experience of infinite perception.

This is a book about what role psychedelics and their ability to open us up to this infinite perception, to allow us to escape the cave, that grant us owl vision in the darkest night, will have in helping us reimagine, reinvent, human civilization at the precisely the moment when it has become completely globalized, yet is poised at the brink of ecological collapse, world war, widespread poverty and inequality, and massive mental and spiritual pain.

This last bit is possibly what is driving the search for "other possibilities" harder than any of the others. Our own pain.

While each chapter in this book can be simply read on its own and will offer the reader a glimpse into one of these "other possibilities," the book is also organized to have a continuous thematic arc from beginning to end. It is a journey in and of itself that opens with a cosmic vision and closes with a personal one, yet at the same time, shows that those are the same thing. That is the power of infinite perception.

Each chapter is a glimpse into the potential of psychedelic plant medicines to play an active role in global transformation. Together, the chapters of this book help bring some clarity to just what exactly we are doing, what exactly we are aiming at, when we talk of a psychedelic renaissance and the role it may have in transforming our world into something more just, sustainable, and sacred. The mystical experience lies at the heart of this, and it is the continuous theme that weaves its way through this book.

The first section, "Cultural Transformation," begins with indigenous perspectives that ground us in the true use of visionary medicine. We are to harmonize with the world, reconnect to nature and understand that we are right now at a point of spiritual evolution. We then move into chapters that address some of our core cultural dilemmas that stand in the way of that evolution, from Zoe Helene showing how psychedelics can help us deconstruct the entrenched patriarchy that has defined male as dominator over not just female but the entire natural world, to Chaikuni Witan on how they can help us heal from the plague of racism, which has institutionalized the opposite of brotherly love.

The next section, "Societal Transformation," dives right into perspective changes at the societal level, from how psychedelics could help us transform the medical, legal and penal systems to how the psychedelic movement risks being co-opted by those very same entrenched institutions. This section includes entries from some of the top organizations involved in psychedelic research and their integration into modern society including MAPS, ICEERS, and Decriminalize Nature. A chapter on ayahuasca as a tool of resistance for indigenous Amazonian cultures, which has many deep implications for spiritually grounded revolutionary movements in the West, is also included here.

The last section zooms in to "Personal Transformation" itself. Your own experience is the last word after all, and it is here that you can read personal stories of those that have worked with psychedelic plant medicines in ways that have brought about healing and transformation. This section also includes several voices who have devoted their lives to working with specific plant medicines including author Elizabeth Bast, who shares the wisdom she has gained through apprenticing with the Bwiti tradition and the sacred Iboga plant, and ethnobotanist Scott Lite, who has been rediscovering the lost traditions of the Huachuma cactus.

We are publishing this book because in the process of global transformation, psychedelics will undoubtedly play a key role. While they are not the only doorway to the mystic experience, they are indeed a fast pass. A blinding flash that illuminates deeper levels of reality. In the right context and conditions, the right set and setting, with the right intentions and guidance, just like many indigenous and traditional societies have been using them for thousands of years, they are powerful tools for reharmonizing human consciousness with the true fundamental laws of the cosmos and the awe-inspiring complexity and abundance of Mother Nature.

It makes sense that the mental health crisis, the massive amount of spiritual pain in the world right now, is driving the new "psychedelic renaissance." But let's get to the root of it, not just treat the symptoms. After all, personal healing and global transformation go hand in hand, are infinitely connected. A world without war and environmental degradation, without racism and sexism, and without monotonous mechanized lifestyles, would also be a world without depression, addiction, and other mental illnesses.

Only wisdom drawn from the mystical realm, long suppressed in Western civilization, but still intimately part of everyday life for indigenous societies, will bring us out of the cave. Psychedelics, with their ability to catalyze a mystic experience in each and every one of us, are a tool, a torch, to help us find the way.

May this book inspire you to join in on this incredible journey as globalized humanity awakes to its true potential and dreams the world anew.

Ocean Malandra

References

1. Kangaslampi, S. (2023). "Association between mystical-type experiences under psychedelics and improvements in well-being or mental health – A comprehensive review of the evidence." *Journal of Psychedelic Studies*, 7(1), 18–28.

2. Sigdyal, Pratigya; Najmi, Hossein; Boedeker, Peter. "Student-Loan and Depression: Implications for Higher Education Educators." *Research in Higher Education Journal*, v. 41 May 2022.

3. Hans S. Schroder, Andrew Devendorf, Brian J. Zikmund-Fisher. "Framing depression as a functional signal, not a disease: Rationale and initial randomized controlled trial." *Social Science & Medicine*, Volume 328, 2023.

Overview

Transcending the Conditioned Mind: Reprogramming through Psychedelics

Natalie L. Dyer, PhD

Global society has reached a crossroads with an uncertain trajectory. The path toward sophisticated artificial intelligence paired with the corporate media-pharmaceutical-agricultural-military hydra is a dangerous cocktail capable of destroying all life and the planet. People and animals are suffering needlessly, knowingly, and despite having the financial and practical means to end this suffering, we continue to prioritize other, less important things such as entertainment or war. Societal sickness is reflected in the mental health of its citizens, which has not shown any strong signs of improvement. At this critical juncture we need to widen our circle of compassion and see ourselves as the whole of which we are a part. One key to assist the transformation of the human psyche and society may be psychedelics. Psychedelics have a remarkable ability to alter consciousness which often leads to positive transformation mediated by ego dissolution, mystical experiences, and shifting the narratives we have about our life. This transformation is manifested as improved psychological health, greater empathy, life meaning, and positive changes in worldview.

Psychedelics can change perception in such a powerful way as to reprogram our psyche toward greater psychological health, compassion toward others, and stewardship toward nature. To change our psychology, we need to change our way of thinking, our beliefs, our programming. The brain is wired to repeat neural activation based on previous use, so it is easy to get stuck in certain modes of thinking and emotions day to day.

It is as if we create grooves in our mind that make thinking and feeling in different ways quite difficult under normal conditions. Psychedelics disrupt this groove, often leading to new understandings about the nature of reality and our relationship to each other and the whole. This meaningful shift can psychologically transform individuals stuck in suffering, anhedonia, or apathy toward life as life becomes infused with fresh energy and deeper meaning. Psychedelics can radically change the way we look at the world in a single instant. Perception, cognition, and emotion alter in ways that are often ineffable.

Healing the Mind

Psychedelics have been used therapeutically for thousands of years (Hofmann, 1980) and recent estimates suggest that there currently are over 30 million lifetime users of psychedelics in the United States (Krebs & Johansen, 2013). Many of these users are self-medicating with psychedelics for mental health reasons (Carhart-Harris & Nutt, 2010; Carhart-Harris & Goodwin, 2017). Indeed, certain psychedelic substances like psilocybin from magic mushrooms, or LSD, have been scientifically shown to improve psychological health in the short and long-term (Wheeler & Dyer et al., 2020; Carhart-Harris et al., 2016; 2017; Roseman et al., 2017; Watts et al., 2017; Erritzoe et al., 2018). Recent results from clinical trials indicate that some psychedelics have strong psychotherapeutic potential for treating mental health issues (Wheeler & Dyer, 2020) including anxiety (dos Santos et al., 2018), depression (Reiche et al., 2018), post-traumatic stress disorder (PTSD) (Sessa et al., 2019), and substance use disorders (Winkelman, 2014).

In general, psychedelic research comes with stigma and heavy baggage, perhaps because of its dark history involving the CIA, Nazi scientists, and covert mind control experiments (Moreno, 2016), combined with the media spreading fear regarding these substances. There has been concern for decades over the potential association between using psychedelics and developing mental illness. Despite these concerns, large-scale studies of over 130,000 participants revealed no significant relationships between lifetime use of psychedelics and increased mental health problems (Krebs & Johansen, 2013; Johansen & Krebs, 2015). On the contrary, in some cases psychedelic use was associated with fewer mental health problems (Krebs & Johansen, 2013; Johansen & Krebs, 2015).

During the early research with psychedelics, many clinical trials were conducted with favorable results for mental health (Grinspoon & Bakalar, 1979; Krebs & Johansen, 2012; Rucker, Iliff, & Nutt, 2018). The success was such that psychedelic therapy had nearly become part of mainstream psychiatric treatment by the 1960s. However, much of this early psychedelic research had methodological flaws and ethical concerns, with the regulations for conducting research with human subjects not as strict as the present day. For these reasons, combined with fear and its sordid past, the first wave of psychedelic research was met with political backlash and the research was halted and suppressed, preventing further investigation until the 1990s. Since then, there has been a revitalization of psychedelic research comprised of many well-designed, ethically rigorous studies of psychotherapy combined with psychedelics, called psychedelic-assisted psychotherapy (Wheeler & Dyer, 2020; Carhart-Harris & Goodwin, 2017). Clinical trials of psychedelic-assisted psychotherapy are on the rise, allowing psychedelics to regain acceptance as therapeutic agents by both researchers and clinicians (Daly, 2018).

The recent revival of research has been largely focused on the therapeutic potential of psychedelics, with the Multidisciplinary Association for Psychedelic Studies (MAPS) spearheading much of the research (see MAPS founder Rick Doblin's chapter). Different types of psychedelics may be useful for different psychological issues (Wheeler & Dyer, 2020). For example, with respect to the classic serotonergic psychedelics, psilocybin has demonstrated wide-ranging potential for treating anxiety, depression, OCD, and substance use disorders associated with alcohol and nicotine; and LSD has shown preliminary ability to treat symptoms of anxiety associated with terminal illness. The empathogen-entactogen MDMA demonstrates strong potential to treat symptoms of PTSD, and there is some support for its ability to treat anxiety associated with autism. The dissociative ketamine and the indole alkaloid ibogaine have shown preliminary efficacy to treat substance use disorders associated with opioids.

Psychedelic-assisted psychotherapy sessions are often led by licensed professionals who are trained in administering psychedelics and monitoring their use, including how to properly guide the patient to minimize distress and support integration after the experience (Mithoefer et al., 2016). Psychedelics may alter consciousness to provide a deeper healing experience during psychotherapy sessions compared to what may be accessible under a normal state of consciousness (Bogenschutz et al., 2018; Nielson et al., 2018; Mithoefer et al., 2018; Belser et al., 2017; Swift et al., 2017; Wagner et al., 2017; Noorani et al., 2018). Psychedelic-assisted psychotherapy can alleviate symptoms of mental illness for some people who do not find relief with standard treatment options (Roseman et al., 2017; Johnson et al., 2014; Watts et al., 2017; Ot'alora et al., 2018; Bouso et al., 2008). For example, up to one-third of individuals with major depressive disorder are resistant to conventional treatment such as selective serotonin reuptake inhibitors (SSRIs) and psychotherapy. For these patients, psychedelic-assisted psychotherapy may hold promise as an effective treatment for their symptoms (Schenberg, 2018).

Along with improvements in clinical indicators of mental illness, reports from interviews with patients also demonstrate the positive impact of psychedelic-assisted therapy on mental health and allow for a deeper investigation into the possible mediators of their therapeutic benefit. Patients that underwent psychedelic-assisted psychotherapy often report: an acceptance of suppressed emotions, increased connectedness to others, greater forgiveness and compassion towards others and the self, enhanced introspective awareness and understanding, increased motivation and commitment to change, positive changes in worldview, acceptance of death for those with terminal illness, ego dissolution, and profound mystical experiences (Mithoefer et al., 2011; Belser et al., 2017; Swift et al., 2017; Wagner et al., 2017; Watts et al., 2017; Bogenschutz et al., 2018; Mithoefer et al., 2018; Nielson et al., 2018; Noorani et al., 2018).

Making the Unconscious Conscious

Psychedelics allow for aspects of our unconscious minds to become conscious, enabling us to identify and modify our hidden programming; our beliefs and ways of being in the world. How we have been behaving, thinking, and feeling can be perceived more objectively, as we become a detached witness. Because of this, psychedelics can expose our programming in an instant, allowing for the opportunity to reprogram beliefs and behaviors that no longer serve our best interests. During psychedelic-assisted psychotherapy, patients can have access to long-held intense or repressed emotions such as grief, sadness, and traumatic pain, and by confronting these repressed emotions, they may be able to move on and leave their emotional pain in the past (Watts et al., 2017). Interviews with patients following psychedelic psychotherapy indicate that psychedelics seem to encourage connection, acceptance, and dealing with emotions (Moreno, Wiegand, Taitano, & Delgado, 2006). On the other hand, patients report that medications and some short-term talk therapies tended to reinforce their sense of disconnection and avoidance while blunting emotional responses and emotions.

For example, psilocybin-assisted psychotherapy seems to have the opposite action from SSRIs and allowed patients to confront their emotions for processing rather than suppressing them (Moreno et al., 2006). Psychedelics have therapeutic potential through bringing our unconscious shadows to the light of awareness to be transformed.

Psychedelic-Enhanced Intuition: Seeing in Unseen

For tens of thousands of years, humanity's indigenous populations frequently consumed psychoactive plant or fungi medicine for intuitive insight and healing (Hofmann, 1980). Many shamans and medicine men and women use psychedelics for healing by entering their own unconscious mind and the unconscious mind of the one being healed. In this way they alter their consciousness to perceive the hidden aspects of the mind that may be causing suffering for the individual. In some forms of shamanism, depression is akin to the concept of soul loss, in which part of our essence is no longer enlivening our form, often due to trauma. This is when aspects of our conscious self migrate into the unconscious self. Shamans navigate the unconscious realm with the help of tools that shift consciousness to bring aspects that need healing to the light, or the conscious mind. This process brings more spirit or life force back into the body. In energy medicine, we are healthy when we have an even steady flow of energy in, out, and throughout the body. In these traditions, health is related to full embodiment of mind and spirit.

Carl Jung said intuition is "perception via the unconscious" (Jung, 1971). The psychedelic experience involves perceiving much more than we do during the normal waking state. As Aldous Huxley would agree, it opens the reducing valve to allow for greater access to Mind At Large (Huxley, 1954). This is related to the filter theory, which states that the brain acts as a filter for consciousness. This filtering mechanism can be altered, allowing an experience of other aspects of reality not normally perceived. This alteration may also be occurring during mystical experiences, mediumship, and other psychical phenomena. We know that the brain is constantly filtering information, much of which never reaches our awareness. Information from both our external environment and internal mental environment is filtered, meaning we are only ever aware of a very small portion of reality at any given moment. If we think of the brain as a filter for consciousness, we can change the dial and tap into information normally outside of our own internal database. This is a theory for how psi abilities work. In line with this hypothesis, scientific evidence supports the accuracy of some extrasensory perceptions during psychedelic experiences (Luke, 2008). Taking psychedelics can be like opening an inner portal whereby the contents of the unconscious or larger mind begin to flood into the conscious or smaller mind. It can be difficult to close this portal after the psychedelic has worn off, and many can feel as if they have gone crazy. Post-treatment psychological integration sessions are important so that the individual can make sense of what emotions and thoughts were uncovered during the session. Without proper post-session integration, a more serious psychological condition could develop (Cohen et al., 1958; Belser et al., 2017).

Ego Dissolution: Deprogramming the Psyche

Sometimes, psychedelics can permit an extreme experience of disidentification from our usual sense of self, as in the case of ego dissolution or death. Ego dissolution is essentially a disintegration of a subjective identity or self (Letheby & Gerrans, 2017), the experience of disidentifying as a separate subject and a merging of one's consciousness with a greater whole (Milliere, 2017). In this state, the narrative about one's life experience, and habitual patterns of thoughts and behaviors are disrupted, allowing for a novel, more objective perspective of one's situation. There is no longer an identification with one's job, name, friends, family, as if a mask was taken off or a play abruptly ended. This can be a blissful state or divine ecstasy, but it can also bring up tremendous fear and a sensation that one is actually dying. Ego dissolution feels like dying because the false self or persona begins to vanish as the point of identification. We identify as something greater, no longer constrained by the ideas of "little me." The mental vehicle we inhabited to experience earth as a separate human being is no longer in control. We are set free from mental limitations and perceive ourselves as vaster than we do during normal waking consciousness. Therefore, ego dissolution may allow for amelioration of pathological thoughts and behaviors that accompany and contribute to mental illness, and in turn, result in a restructuring of perspective, beliefs, and worldview that leads to positive clinical outcomes. This reconstructive process can be guided and supported by the accompaniment of psychotherapy. Under psychotherapeutic conditions, during ego dissolution the patient can perceive their situation from a different, more objective perspective, enabling them to modify patterns of behavior, thoughts, and emotions that have contributed to their mental illness (Swanson, 2018). The psychedelic therapist will guide or support the patient through this process to promote healing.

Ego dissolution is an aspect of many spiritual practices and traditions whereby the little self merges with the ultimate Self, or God. Along with meditation, yoga, or self-inquiry, psychedelics is a tool for this spiritual discovery or self-realization. Remarkably, the psychedelic experience has many things in common with near death experiences (NDEs) (Timmermann et al., 2018) and other mystical experiences. A recent placebo-controlled study with 13 participants that took N, N-Dimethyltryptamine (DMT) revealed that the DMT experience met criteria for an NDE (Timmermann et al., 2018). Taking DMT was also associated with mystical experiences and ego dissolution (Timmermann et al., 2018). Experiencing something analogous to death on psychedelics may help reduce the fear of dying by becoming acquainted with our impermanence. Research with people with terminal illness seems to support this contention, whereby LSD or psilocybin helps reduce death anxiety (Grob et al., 2011; Gasser et al., 2014a; 2014b; Swift et al., 2017; Malone et al., 2018). Therefore, there are profound implications for the use of psychedelics for improving palliative care. Considering everyone must face their death at some point, this is an important area of research.

The psychological and neurobiological mechanisms for the therapeutic benefits of psychedelics have yet to be fully elucidated. However, preliminary results suggest that one therapeutic mechanism may be through the psychedelic's ability to elicit ego dissolution (Swanson, 2018). As mentioned previously, ego dissolution was a common theme from the qualitative reports of patients' experience while undergoing psychedelic-assisted psychotherapy. Ego dissolution has been shown to occur in a dose-dependent fashion, with lower doses leading to a softening of the ego and higher doses leading to complete ego dissolution (Letheby & Gerrans, 2017; Milliere, 2017).

Spirituality and Meaning-Making

An experience of interconnection with others and the whole is a key part of the psychedelic experience (Carhart-Harris et al., 2018). Beyond the ego, the sense of a separate self, is unity or oneness. There is a realization of being unbound by our physical bodies, untethered to the usual constraints of time and space. The experience often feels more real than normal waking reality, as an expansion of consciousness. Some experience a feeling of unconditional love and peace as a fundamental aspect of reality, behind the scenes of the play of ego. The stereotypical peace and love hippie comes to mind. Feelings of universal love are commonly experienced during psychedelic experiences, and other mystical experiences such as during NDEs, energy healing sessions, and meditation. With universal love, all beings are wished well with deep compassion and reverence for existence (Trent et al., 2019). It makes sense, then, that psychedelics lead to more compassionate, empathic behavior and can lead to reduced violence (Walsh et al., 2016).

Given their potent nature, it is understandable that psychedelics facilitate psychological and spiritual transformation (Wolfson et al., 2011). Wolfson defines transformation as "A change in one's core conceptual and even physical structure that interrupts the prior sense of self and induces an altered, at least partially different sense of self immediately and/or over time with some degree of persistence. Transformation is a reset of the old software with at least some new programming" (p. 982). As evidence for their transformative power, psychedelics are associated with finding greater meaning in life, which is one of the mediators for their therapeutic benefit (Hartogsohn, 2018). Psychedelics can orient the nonbeliever toward enhanced spirituality and greater life meaning. For example, people who had never used psychedelics before reported long-term increases in the belief that there is continuity after death, a transition to something even greater than this life (Griffiths et al., 2011). Moreover, use of psychedelics with a spiritually inclined attitude may act as a protective factor against drug-related problems (Móró et al., 2011), further highlighting the therapeutic interaction between psychedelics and spirituality.

Along with ego dissolution, the meaning derived from the psychedelic experience is also one of the mediators of its therapeutic effect. Research indicates that having a peak mystical experience during psychedelic use is associated with more positive therapeutic outcomes (Pahnke et al., 1969; Richards,

1978; Richards et al., 1980; Krupitsky et al., 2002; Bogenschutz et al., 2015; Garcia-Romeu et al., 2014; Griffiths et al., 2016; Ross et al., 2016; Mithoefer et al., 2018). Psychedelic psychotherapists often encourage patients to let go of feelings of control in an attempt to facilitate peak psychedelic experiences that may provide greater psychotherapeutic benefits.

Neuroscience of Ego Death

Reprogramming our self ultimately means rewiring our brain through changing our beliefs, thoughts, and patterns of behavior. The reprogramming effect of psychedelics can be observed in the brain. Recent research suggests that psychedelic drugs perturb brain networks and processes that normally constrain neural systems central to perception, emotion, cognition, and sense of self or ego (Swanson, 2018). That means the filter is reduced, expanding perception and the concepts of who we are. One resting state network, the default mode network (DMN), comprised of the medial prefrontal cortex and posterior cingulate cortex, is active during mind wandering and rumination related to the self, and may be the neurobiological seat of the ego (Carhart-Harris & Friston, 2010). Neuroimaging studies have revealed that the integrity of the default mode network is modulated by psychedelics including psilocybin (Carhart-Harris et al., 2012; 2013), LSD (Tagliazucchi et al., 2017), ayahuasca (Palhano-Fontes et al., 2015), ketamine (Scheidegger et al., 2012), and MDMA, though to a lesser degree (Roseman et al., 2014). Furthermore, decreased functional connectivity and oscillatory power of the default mode network was found to be correlated with ego dissolution when under the influence of LSD (Tagliazucchi et al., 2017) and psilocybin (Lebedev et al., 2015). The greater the ego dissolution, the greater the default mode network is deactivated and the greater the therapeutic effect (Tagliazucchi et al., 2016). These studies may support the filter theory of consciousness described earlier, whereby the normally constrained system, the filter, becomes less constrained and consciousness is expanded to encompass a greater aspect of reality (Swanson, 2018).

The neuroimaging data supports the hypothesis that ego dissolution is at least one mechanism for the therapeutic benefit of some psychedelics. Psychedelics seem to decrease use of the default mode network and activate other networks for thinking "outside the box." Psychedelics result in more communication distributed across the brain. The brain becomes more child-like, and essentially enters a reprogrammable state. This enables a sort of return to innocence, leading to seeing painful situations from different perspectives, fostering forgiveness and empathy. Indeed, psychedelics such as LSD and psilocybin have been shown to increase empathy after use (Dolder et al., 2016; Mason et al., 2019).

Conclusions

Psychedelics are powerful tools for psychological transformation through their ability to promote psychological health, ego dissolution, and positively shift worldview toward greater compassion and meaning. Recent scientific research indicates that psychedelic-assisted psychotherapy can be safely used to treat a range of mental health conditions, including anxiety, depression, PTSD, and substance use disorders. Patients report benefits associated with acceptance of emotions, positive changes in the perspective of their illness, and enhanced connectedness to others. The therapeutic benefits of psychedelics may be through their ability to promote ego dissolution and correspondingly decreased integrity of the default mode network (DMN). While psychedelic-assisted psychotherapy research is still at an early stage, the results are promising. More rigorous studies with larger sample sizes should be conducted to determine the most appropriate psychedelic dose, and psychotherapy for different mental health conditions, as well as to further elucidate the psychological and neurobiological mechanisms underlying their therapeutic benefit. Scientific research indicates what we knew all along, that psychedelics are powerfully transformative. As such, they should be respected and used in a respectful, therapeutic manner. Proper psychological integration of the experience should be insured to prevent any negative aftereffects and to consolidate the psychological transformation. For this reason and others, legal psychedelic use with a licensed psychotherapist or highly experienced and qualified medicinal guide is recommended. Psychedelic-like experiences can also be accessed under nonpharmacological conditions, such as during breathing exercises, meditation, or yoga. Use of these practices is also encouraged, as they have wide-ranging benefits to physical, mental, and spiritual health.

Psychedelic substances are powerful tools for transformation and should not be taken simply for recreational reasons. The importance of respect and intention for the experience cannot be overstated. Having said that, it is an exciting time where we are moving toward understanding how to work with plant medicines in a way that is informed by both indigenous wisdom and contemporary research. We are no longer existing in separate villages, dependent on a few for survival and social health, but connected across the planet in a myriad of ways. To move forward positively as a species, we must shift our consciousness into a higher perspective and think in terms of the whole as a global culture. Psychedelics may be one way in which we can prevent global catastrophe through positively transforming the human psyche toward greater psychological health and an awareness of our interconnection.

References

Belser, A.B., Agin-Liebes, G., Swift, T.C., Terrana, S., Devenot, N., Friedman, H.L. et al. (2017). Patient experiences of psilocybin-assisted psychotherapy: an interpretative phenomenological analysis. *Journal of Humanistic Psychology*, 57(4), 354–388. doi:10.1177/0022167817706884.

Bogenschutz, M.P., Forcehimes, A.A., Pommy, J.A., Wilcox, C.E., Barbosa, P., & Strassman, R.J. (2015). Psilocybin-assisted treatment for alcohol dependence: A proof-of-concept study. *Journal of Psychopharmacology*, 29(3), 289–299. doi:10.1177/0269881114565144.

Bogenschutz, M.P., Podrebarac, S.K., Duane, J.H., Amegadzie, S.S., Malone, T.C., Owens, L.T. et al. (2018). Clinical interpretations of patient experience in a trial of psilocybin-assisted psychotherapy for alcohol use disorder. *Frontiers in Pharmacology*, 9, 100. doi:10.3389/fphar.2018.00100.

Bouso, J.C., Doblin, R., Farré, M., Alcázar, M.Á., & Gómez-Jarabo, G. (2008). MDMA-assisted psychotherapy using low doses in a small sample of women with chronic posttraumatic stress disorder. *Journal of Psychoactive Drugs*, 40(3), 225–236. doi:10.1080/02791072.2008.10400637.

Carhart-Harris, R.L., & Nutt, D.J. (2010). User perceptions of the benefits and harms of hallucinogenic drug use: A web-based questionnaire study. *Journal of Substance Use*, 15(4), 283–300.

Carhart-Harris, R.L., & Friston, K.J. (2010). The default-mode, ego-functions and free-energy: a neurobiological account of Freudian ideas. *Brain*, 133(4), 1265–1283.

Carhart-Harris, R.L., Erritzoe, D., Williams, T., Stone, J.M., Reed, L.J., Colasanti, A. et al. (2012). Neural correlates of the psychedelic state as determined by fMRI studies with psilocybin. *Proceedings of the National Academy of Sciences*, 109(6), 2138–2143. doi:10.1073/pnas.1119598109.

Carhart-Harris, R.L., Leech, R., Erritzoe, D., Williams, T.M., Stone, J.M., Evans, J. et al. (2012). Functional connectivity measures after psilocybin inform a novel hypothesis of early psychosis. *Schizophrenia Bulletin*, 39(6), 1343–1351. doi:10.1093/schbul/sbs117.

Carhart-Harris, R. (2013). Psychedelic drugs, magical thinking and psychosis. *Journal of Neurology, Neurosurgery & Psychiatry*, 84(9). doi:10.1136/jnnp-2013-306103.17.

Carhart-Harris, R.L., Leech, R., Hellyer, P.J., Shanahan, M., Feilding, A., Tagliazucchi, E. et al. (2014). The entropic brain: A theory of conscious states informed by neuroimaging research with psychedelic drugs. *Frontiers in Human Neuroscience*, 8. doi:10.3389/fnhum.2014.00020.

Carhart-Harris, R.L., Bolstridge, M., Rucker, J., Day, C.M.J., Erritzoe, D., Kaelen, M. et al. (2016). Psilocybin with psychological support for treatment-resistant depression: an open-label feasibility study. *Lancet Psychiatry*, 3(7), 619–627.

Carhart-Harris, R.L., Bolstridge, M., Day, C.M., Rucker, J., Watts, R., Erritzoe, D.E. et al. (2017). Psilocybin with psychological support for treatment-resistant depression: Six-month follow-up. *Psychopharmacology*, 235(2), 399–408. doi:10.1007/s00213-017-4771-x.

Carhart-Harris, R.L., & Goodwin, G.M. (2017). The therapeutic potential of psychedelic drugs: past, present, and future. *Neuropsychopharmacology*, 42(11), 2105.

Carhart-Harris, R.L., Erritzoe, D., Haijen, E., Kaelen, M., & Watts, R. (2018). Psychedelics and connectedness. *Psychopharmacology*, 235(2), 547–550.

Cohen, S., Fichman, L., & Eisner, B.G. (1958). Subjective reports of lysergic acid experiences in a context of psychological test performance. *American Journal of Psychiatry*, 115(1), 30–35.

Daly, M. (2018). What can we learn from the psychedelic renaissance? *Matters of Substance*, 29(1), 22.

Dolder, P.C., Schmid, Y., Müller, F., Borgwardt, S., & Liechti, M.E. (2016). LSD acutely impairs fear recognition and enhances emotional empathy and sociality. *Neuropsychopharmacology*, 41(11), 2638.

dos Santos, R.G., Osório, F.L., Crippa, J.A.S., Bouso, J.C., & Hallak, J.E.C. (2018). The therapeutic potential of ayahuasca and other serotonergic hallucinogens in the treatment of social anxiety. *Social Anxiety Disorder: Recognition, Diagnosis and Management*, 183–206.

Dossey, L. (2013). *One Mind*. Hay House Publishing.

Erritzoe, D., Roseman, L., Nour, M.M., Maclean, K., Kaelen, M., Nutt, D.J., & Carhart-Harris, R.L. (2018). Effects of psilocybin therapy on personality structure. *Acta Psychiatrica Scandinavica*, 138(5), 368–378. doi:10.1111/acps.12904.

Garcia-Romeu, A., Griffiths, R.R., & Johnson, M.W. (2014). Psilocybin-occasioned mystical experiences in the treatment of tobacco addiction. *Current Drug Abuse Reviews*, 7(3), 157–164.

Gasser, P., Holstein, D., Michel, Y., Doblin, R., Yazar-Klosinski, B., Passie, T., & Brenneisen, R. (2014a). Safety and efficacy of lysergic acid diethylamide-assisted psychotherapy for anxiety associated with life-threatening diseases. *The Journal of Nervous and Mental Disease*, 202(7), 513–520. doi:10.1097/nmd.0000000000000113.

Gasser, P., Kirchner, K., & Passie, T. (2014b). LSD-assisted psychotherapy for anxiety associated with a life-threatening disease: A qualitative study of acute and sustained subjective effects. *Journal of Psychopharmacology*, 29(1), 57–68. doi:10.1177/0269881114555249.

Griffiths, R.R., Johnson, M.W., Richards, W.A., Richards, B.D., McCann, U., Jesse, R. (2011). Psilocybin occasioned mystical-type experiences: Immediate and persisting dose-related effects. *Psychopharmacology* (Berl), 218, 649–665.

Grinspoon, L., Bakalar, J.B. (1979). *Psychedelic Drugs Reconsidered*. New York: Basic Books.

Grob, C.S., Danforth, A.L., Chopra, G.S., Hagerty, M., McKay, C.R., Halberstadt, A.L., & Greer, G.R. (2011). Pilot study of psilocybin treatment for anxiety in patients with advanced-stage cancer. *Archives of General Psychiatry*, 68(1), 71–78. doi:10.1001/archgenpsychiatry.2010.116.

Hartogsohn, I. (2018). The meaning-enhancing properties of psychedelics and their mediator role in psychedelic therapy, spirituality, and creativity. *Frontiers in Neuroscience*, 12, 129.

Hofmann, A. (1980). *LSD: My problem child*. New York: McGraw-Hill.

Huxley, A. (1954). *The Doors of Perception* and *Heaven and Hell*. London: Flamingo.

Johansen, P.Ø., & Krebs, T.S. (2015). Psychedelics not linked to mental health problems or suicidal behavior: A population study. *Journal of Psychopharmacology*, 29(3), 270–279.

Johnson, M.W., Garcia-Romeu, A., Cosimano, M.P., & Griffiths, R.R. (2014). Pilot study of the 5-HT$_{2A}$R agonist psilocybin in the treatment of tobacco addiction. *Journal of Psychopharmacology*, 28(11), 983–992. doi:10.1177/0269881114548296.

Jung, C.G. (1971). *Psychological Types*. Bollingen Series XX, Volume 6, Princeton University Press.

Krebs, T.S., & Johansen, P.Ø. (2012). Lysergic acid diethylamide (LSD) for alcoholism: Meta-analysis of randomized controlled trials. *Journal of Psychopharmacology*, 26(7), 994–1002. doi:10.1177/0269881112439253.

Krebs, T.S., & Johansen, P.Ø. (2013). Psychedelics and mental health: a population study. *PLoS One*, 8(8), e63972.

Krupitsky, E., Burakov, A., Romanova, T., Dunaevsky, I., Strassman, R., & Grinenko, A. (2002). Ketamine psychotherapy for heroin addiction: Immediate effects and two-year follow-up. *Journal of Substance Abuse Treatment*, 23(4), 273–283. doi:10.1016/s0740-5472(02)00275-1.

Lebedev, A.V., Lövdén, M., Rosenthal, G., Feilding, A., Nutt, D.J., & Carhart-Harris, R.L. (2015). Finding the self by losing the self: Neural correlates of ego-dissolution under psilocybin. *Human Brain Mapping*, 36(8), 3137–3153.

Letheby, C., & Gerrans, P. (2017). Self unbound: Ego dissolution in psychedelic experience. *Neuroscience of Consciousness*, 1. doi:10.1093/nc/nix016.

Luke, D.P. (2008). Psychedelic Substances and Paranormal Phenomena: A Review of the Research. *Journal of Parapsychology*, 72(1).

Malone, T.C., Mennenga, S.E., Guss, J., Podrebarac, S.K., Owens, L.T., Bossis, A.P. et al. (2018). Individual experiences in four cancer patients following psilocybin-assisted psychotherapy. *Frontiers in Pharmacology*, 9, 256. doi:10.3389/fphar.2018.00256.

Mason, N.L., Mischler, E., Uthaug, M.V., & Kuypers, K.P. (2019). Sub-acute effects of psilocybin on empathy, creative thinking, and subjective well-being. *Journal of Psychoactive Drugs*, 51(2), 123–134.

Milliere, R. (2017). Looking for the self: phenomenology, neurophysiology and philosophical significance of drug-induced ego dissolution. *Frontiers in Human Neuroscience*, 11, 245.

Mithoefer, M. et al. (2016). A manual for MDMA-assisted therapy in the treatment of PTSD; Version 8. http://www. maps.org/research/mdma/mdma-research-timeline/4887-a-manual-for-mdma-assisted-psychotherapy-in-the-treatment-of-ptsd

Mithoefer, M.C., Mithoefer, A.T., Feduccia, A.A., Jerome, L., Wagner, M., Wymer, J. et al. (2018). 3, 4-methylenedioxymethamphetamine (MDMA)-assisted psychotherapy for post-traumatic stress disorder in military veterans, firefighters, and police officers: A randomised, double-blind, dose-response, phase 2 clinical trial. *Lancet Psychiatry*, 5(6), 486–497. http://dx.doi.org.proxy1.lib.uwo.ca/10.1016/S2215-0366(18)30135-4

Moreno, F.A., Wiegand, C.B., Taitano, E.K., & Delgado, P.L. (2006). Safety, tolerability, and efficacy of psilocybin in 9 patients with obsessive-compulsive disorder. *The Journal of Clinical Psychiatry*, 67(11), 1735–1740. doi:10.4088/jcp.v67n1110.

Moreno, J.D. (2016). Acid Brothers: Henry Beecher, Timothy Leary, and the psychedelic of the century. *Perspectives in Biology and Medicine*, 59(1), 107–121.

Móró, L., Simon, K., Bárd, I., & Racz, J. (2011). Voice of the psychonauts: Coping, life purpose, and spirituality in psychedelic drug users. *Journal of Psychoactive Drugs*, 43(3), 188–198.

Nielson, E.M., May, D.G., Forcehimes, A.A., & Bogenschutz, M.P. (2018). The psychedelic debriefing in alcohol dependence treatment: Illustrating key change phenomena through qualitative content analysis of clinical sessions. *Frontiers in Pharmacology*, 9, 132. doi:10.3389/fphar.2018.00132.

Noorani, T., Garcia-Romeu, A., Swift, T.C., Griffiths, R.R., & Johnson, M.W. (2018). Psychedelic therapy for smoking cessation: Qualitative analysis of participant accounts. *Journal of Psychopharmacology*, 32(7), 756–769. doi:10.1177/0269881118780612.

Nummenmaa, L., Glerean, E., Hari, R., & Hietanen, J.K. (2014). Bodily maps of emotions. *Proceedings of the National Academy of Sciences*, 111(2), 646–651.

Ot'alora, M., Grigsby, J., Poulter, B., Van Derveer, J.W., Giron, S.G., Jerome, L. et al. (2018). 3, 4-Methylenedioxymethamphetamine-assisted psychotherapy for treatment of chronic posttraumatic stress disorder: A randomized phase 2 controlled trial. *Journal of Psychopharmacology*, 32(12), 1295–1307. doi:10.1177/0269881118806297.

Pahnke, W.N. (1969). The psychedelic mystical experience in the human encounter with death. *Harvard Theological Review*, 62(01), 1–21. doi:10.1017/s0017816000027577.

Palhano-Fontes, F., Andrade, K.C., Tofoli, L.F., Santos, A.C., Crippa, J.A.S., Hallak, J.E.C. et al. (2015). The psychedelic state induced by ayahuasca modulates the activity and connectivity of the default mode network. *PLoS One*, 10(2), e0118143.

Reiche, S., Hermle, L., Gutwinski, S., Jungaberle, H., Gasser, P., & Majić, T. (2018). Serotonergic hallucinogens in the treatment of anxiety and depression in patients suffering from a life-threatening disease: A systematic review. *Progress in Neuropsychopharmacology and Biological Psychiatry*, 81, 1–10. doi:10.1016/j.pnpbp.2017.09.012.

Richards, W.A. (1978). Mystical and archetypal experiences of terminal patients in DPT-assisted psychotherapy. *Journal of Religion & Health*, 17(2), 117–126. doi:10.1007/bf01532413.

Richards, W.A. (1980). Psychedelic drug-assisted psychotherapy with persons suffering from terminal cancer. *Journal of Altered States of Consciousness*, 5, 309–319.

Roseman, L., Leech, R., Feilding, A., Nutt, D.J., & Carhart-Harris, R.L. (2014). The effects of psilocybin and MDMA on between-network resting state functional connectivity in healthy volunteers. *Frontiers in Human Neuroscience*, 8. doi:10.3389/fnhum.2014.00204.

Roseman, L., Demetriou, L., Wall, M.B., Nutt, D.J., & Carhart-Harris, R.L. (2017). Increased amygdala responses to emotional faces after psilocybin for treatment-resistant depression. *Neuropharmacology*, 142, 263–269. doi:10.1016/j.neuropharm.2017.12.041.

Ross, S., Bossis, A., Guss, J., Agin-Liebes, G., Malone, T., Cohen, B. et al. (2016). Rapid and sustained symptom reduction following psilocybin treatment for anxiety and depression in patients with life-threatening cancer: A randomized controlled trial. *Journal of Psychopharmacology*, 30(12), 1165–1180. doi:10.1177/0269881116675512.

Rucker, J.J., Iliff, J., & Nutt, D.J. (2018). Psychiatry & the psychedelic drugs. Past, present & future. *Neuropharmacology*, 142, 200–218.

Scheidegger, M., Walter, M., Lehmann, M., Metzger, C., Grimm, S., Boeker, H. et al. (2012). Ketamine decreases resting state functional network connectivity in healthy subjects:

implications for antidepressant drug action. *PLoS ONE*, 7(9). doi:10.1371/journal.pone.0044799.

Schenberg, E.E. (2018). Psychedelic-assisted psychotherapy: a paradigm shift in psychiatric research and development. *Frontiers in Pharmacology*, 9, 733.

Sessa, B., Higbed, L., & Nutt, D. (2019). A review of 3,4-methylenedioxymethamphetamine (MDMA)-assisted psychotherapy. *Frontiers in Psychiatry*, 10. doi:10.3389/fpsyt.2019.00138.

Swanson, L.R. (2018). Unifying theories of psychedelic drug effects. *Frontiers in Pharmacology*, 9, 172.

Swift, T.C., Belser, A.B., Agin-Liebes, G., Devenot, N., Terrana, S., Friedman, H.L. et al. (2017). Cancer at the dinner table: experiences of psilocybin-assisted psychotherapy for the treatment of cancer-related distress. *Journal of Humanistic Psychology*, 57(5), 488–519.

Tagliazucchi, E., Roseman, L., Kaelen, M., Orban, C., Muthukumaraswamy, S.D., Murphy, K. et al. (2016). Increased global functional connectivity correlates with LSD-induced ego dissolution. *Current Biology*, 26(8), 1043–1050.

Timmermann, C., Roseman, L., Williams, L., Erritzoe, D., Martial, C., Cassol, H. et al. (2018). DMT models the near-death experience. *Frontiers in Psychology*, 9, 1424.

Trent, N.L., Beauregard, M., & Schwartz, G.E. (2019). Preliminary development and validation of a scale to measure universal love. *Spirituality in Clinical Practice*.

Trichter, S., Klimo, J., & Krippner, S. (2009). Changes in spirituality among ayahuasca ceremony novice participants. *Journal of Psychoactive Drugs*, 41(2), 121–134.

Wagner, M.T., Mithoefer, M.C., Mithoefer, A.T., MacAulay, R.K., Jerome, L., Yazar-Klosinski, B., & Doblin, R. (2017). Therapeutic effect of increased openness: Investigating mechanism of action in MDMA-assisted psychotherapy. *Journal of Psychopharmacology*, 31(8), 967–974. doi:10.1177/0269881117711712.

Walsh, Z., Hendricks, P.S., Smith, S., Kosson, D.S., Thiessen, M.S., Lucas, P., & Swogger, M.T. (2016). Hallucinogen use and intimate partner violence: Prospective evidence consistent with protective effects among men with histories of problematic substance use. *Journal of Psychopharmacology,* 30(7), 601–607.

Watts, R., Day, C., Krzanowski, J., Nutt, D., & Carhart-Harris, R. (2017). Patients' accounts of increased "connectedness" and "acceptance" after psilocybin for treatment-resistant depression. *Journal of Humanistic Psychology,* 57(5), 520–564. doi:10.1177/0022167817709585.

Wheeler, S.W., & Dyer, N.L. (2020). A systematic review of psychedelic-assisted psychotherapy for mental health: An evaluation of the current wave of research and suggestions for the future. *Psychology of Consciousness: Theory, Research, and Practice,* 7(3), 279.

Winkelman, M. (2014). Psychedelics as medicines for substance abuse rehabilitation: evaluating treatments with LSD, peyote, ibogaine and ayahuasca. *Current Drug Abuse Reviews,* 7(2), 101–116. doi:10.2174/1874473708666150107120011.

Wolfson, P. (2011). A longitudinal history of self-transformation: Psychedelics, spirituality, activism and transformation. *Journal of Consciousness Exploration & Research,* 2(7).

Section 1

Cultural Transformation

1

You Must Harmonize with the Universe

Interview with Matilde Gomez Silvado, Shipibo Shaman

Matilde Gomez Silvado is a curandera of Shipibo ethnicity that lives in the jungle just outside of Nauta, Peru, with her extended family. She holds regular ayahuasca ceremonies for both locals and visitors and uses a wide variety of different plant medicines for healing. The Shipibo are one of the few indigenous communities in the Amazon basin that still have female shamans, although there is overwhelming evidence that women commonly held leadership positions, both politically and spiritually, before European colonization of the area.

Interview conducted by Ocean Malandra

∞

OM: Good afternoon, Matilde, and thanks for taking the time to talk with me. My first question for you is: what is the role of shamans in the Shipibo community?

Matilde: My grandmother was a great shaman, and so it passes down the family like this. How to deal with problems, do cures and *sanaciones* (cleansings) for people, all of this I learned from her. And then from my grandfather, he taught me about different plants. Later, I dieted different plants in order to learn from them and how to work with them. The medicinal plants give us energy to cure diseases and help other people. A shaman works with ayahuasca and medicinal plants in order to help people.

OM: Why are so many foreigners coming to the Amazon to drink ayahuasca?

Matilde: Foreigners are coming to the jungle because they have serious problems. They have many different problems that are causing them to come to Peru and try to cleanse themselves. I work with lots of females who come because they have chronic problems in their female parts and reproductive systems. We try to treat them with plants like *uvos*, *chuchuwasa* and *una de gato*.

Sometimes they stay for one week, sometimes two. Sometimes they stay for a month, sometimes two. Sometimes eight. Sometimes years.

My grandmother dieted plants for a year straight and became a strong curandero. She is the one who taught me and helped me to become a healer too. Now she is 120 years old.

OM: What do you think the future of humanity looks like? Are we going to evolve to a better way of life here on planet earth? Will ayahuasca help us do this?

Matilde: Ayahuasca can help people change their feelings about things and help them become healthier. This change can make them more conscious people. Conscious people create a better society.

OM: What are your recommendations regarding trying ayahuasca?

Matilde: The people who come here to me come to drink ayahuasca because they want to cleanse and cure themselves. But the curandero also has an important role. We sing songs and "icaros," which support them on an energetic level, and that comes from the plants we work with too.

OM: And what is the real reason for drinking ayahuasca? What is the ultimate goal?

Matilde: Ayahuasca is for having visions that lead towards harmonization with the universe. When real harmony is realized, then you have peace.

2

Consciousness Is the Cure

Interview with Miguel Evanjuanoy of the Union of Traditional Yagé Medics of the Colombian Amazon (UMIYAC)

Miguel Evanjuanoy is the spokesperson for the Union of Traditional Yagé Medics of the Colombian Amazon (UMIYAC), an organization made up of spiritual leaders from five different indigenous ethnicities in Colombia and Ecuador that all use yagé (ayahuasca) traditionally in their communities. Miguel is from the Inga community, whose ancestral territory lies in the lush rainforest-covered mountains of the Putumayo department of Colombia. He is also a board member of the Indigenous Medicine Conservation Fund (IMC), which works to preserve traditional medicine traditions in indigenous communities around the world.

Interview conducted by Ocean Malandra

∞

OM: Miguel. let's start with the basics. Why was UMIYAC formed? What were the issues and problems that the indigenous communities in the Colombian Amazon were facing that made them want to join together and create UMIYAC?

Miguel: That's an important thing to explain. This was a way to unite the pueblos, and in a spiritual way. It was the grandfathers and grandmothers, who have the longest time immersed in the spiritual traditions, that saw the necessity for this. And the young leaders too, we saw that we needed to form an organization like this. This was in 1999.

All the grandparents got together to talk amongst themselves about Unity. So, what was the reason to create this organization? Indigenous people have a long historical memory, and right now we have been resisting for more than 500 years, and we will keep fighting for 500 more if that's what it takes to get back to the *buen vivir* (good life).

We know that the world out there, and it's not an indigenous world, it's the Western world, wants to end the indigenous way of life, and that's for different reasons and interests. Supposedly, it's for progress and development. But this development has us very worried. And we created this organization because we felt very threatened.

Each different community has their spirituality, but we were not connected, and that made us more divided, more fragile. This made us more vulnerable to the strategies of the state, and I'm not just talking about the Colombian government. Here where we live in the Amazon basin, there are many different businesses and interests with extractive goals, that are not from here, that come from different countries all over the world.

So, because we felt threatened, we decided to come together, five different indigenous communities that all employ yagé in our spirituality. It's a spiritual based organization. We know that the invasion, the destruction and extraction of the forest and the territories, the sickness and plagues, and everything else that comes in from the outside world, is part of a strategy to divide us. To generate divisions among the pueblos, the territories, and the people—including infighting and even war between the communities.

These businesses were even buying leaders in the communities, paying them to sow discord among the people so that they could get access to the natural resources of the territories. But all the problems that came with this started to make the people think. And they started to organize. Not just this organization UMIYAC, but different organizations of indigenous communities.

In this case, UMIYAC is a spiritual organization. It's made up exclusively of spiritual leaders that drink yagé. And so we are made up of *yagéceros* from five different indigenous pueblos: Los Inga, el pueblo Siona, el pueblo Koreguaje, el pueblo Kamëntšá, and el pueblo Cofán. And the unifying mission is this: to defend the spirituality, to defend the culture, to defend the autonomy, and to defend the territories from the extractive interests that have come to our lands.

Our organization is a continuation of our resistance, but not just in word, in a spiritual way that helps us preserve unity and stay stronger. Yagé is a sacred plant, a spiritual plant, a teacher plant, a powerful plant that guides us on the path of our indigenous way of life for all the Amazonian communities.

And this is important, because all five of the communities that make up UMIYAC have been declared in risk of cultural and physical extermination by the Colombian Constitution Court. So ultimately, we formed as a spiritual defense against our own extinction.

This is the message that we have for the world: in this time of great changes and great problems, we must keep guarding and defending life, we must keep guarding and protecting Mother Earth. Our Mother Earth is very wounded, she is very hurt, and is very tired. But in spite of this, she does not stop producing the foods and medicines that we need to survive.

So we must not stop protecting and defending her and all the treasures she produces for us out of the fertile earth.

OM: And so you five communities that make up UMIYAC all use yagé as part of your spirituality? What I would like to know is what exactly is the role of this plant medicine in the communities? How does it help you?

Miguel: Thank you for that question. It's very important to talk about the spirituality of the indigenous communities of the Amazon. And the Amazon is our home, not just in Colombia, but we have members who live in Ecuador as well. We are in the departments of Putumayo, Cauca, Caquetá, and in our brother country of Ecuador. All of this is Amazon.

When we talk about our spirituality, we are talking about thoughts passed down from our grandparents, and so this is not easy to explain. But that is the responsibility that our leaders carry. Our spirituality is based in the sacred plant yagé. And we are talking about spiritual traditions with this plant that go back millennia.

If I have a little bit of consciousness of this spirituality, it's because I have shared with the grandparents and they have passed it down to me. And we also share our thoughts, our consciousness, our spirituality, between communities.

But before we share our spirituality with the outside world, we have some worries. Worries about what is happening on a planetary level. Especially what we hear happening with our medicine, the disorder that is taking place related to our sacred medicine.

We see yagé as a spiritual tool that helps us to maintain the protection of our territories and our culture, not as a business, or something that can be bought and sold on a market. There are many people who are blind to this in the outside world, in Europe, in the United States, and in both Central and South America as well.

When they talk of the "ayahuasca," they are only talking about the material aspect. Not the immaterial. The spiritual. We have heard many people from the outside, people who study things in laboratories, say that yagé contains DMT, that it's a psychedelic substance. They say they are investigating, they are discovering. But discovering what? Indigenous communities have been using yagé for thousands of years.

The problem is that many people think just drinking yagé, drinking ayahuasca, that is everything. That with drinking it, they now know it. But all they know is the cup. Not what it contains. They don't know its real story. The scientists, those that study things in laboratories, they think that just drinking it and simply perceiving what the effects are, that is everything. But this is blind.

And this also creates damage. I have been going to these different conferences and events all over the world. And this is to share a message. Not so much to talk about the spirituality of yagé but to talk about what is going on with the planet. For us, Mother Earth is our primary focus, but we do not see this taking place with all these scientists and conferences about ayahuasca and other plants.

So, what is yagé for us? How is our spirituality based on this sacred plant? We have many people who visit us, and they talk about things like laws that are being passed around the world, about different religions. But us, we begin with one thing. We begin with the connection.

Spiritual connection. The connection with Mother Earth. The connection with the territory and the connection with everything that exists on the entire planet. Water, fire, wind and

Mother Earth, everything that moves, and everything that does not move. And there is the difference.

When the Western world arrived here they called us savages. They said we did not know anything. That we were crazy even. But all this is a lie. The crazy ones were them. The savages were them. Because we knew how to live the good life, to coexist.

So this is what we are really talking about when we speak of the spiritual. Many people have lost the real spiritual connection. People do not feel connected to the territories where they live, they are confined to their four walls only. They are not in harmony with the larger world.

When we talk about yagé here in the Amazon, we are talking about this connection. We are talking about this harmony. We drink yagé in our traditional community houses, that many call malocas, but every territory has them. This is where we drink yagé with our grandparents, those who have taken responsibility for the community and the territory, and share their wisdom and perspectives with the community in general.

We do not drink yagé just to cure something. We drink yagé to live connected, to harmonize with the spiritual world, with Mother Earth. We reconnect with the principles of Mother Earth. We respect the connection between the community and the territory so that we can live harmonized. We drink for the protection of the leaders, so that the community lives collectively, so that the leaders are guided in the right way. We drink together so as not to be divided, and therefore fragile. We drink to not fall into the trap of corruption.

For all of this we drink. Not just to be cured in the physical sense. But in the spiritual, which means to live in harmony with the territory and with Mother Earth. We drink to see our errors, where we have failed in protecting Mother Earth.

This is the big difference between what's happening in the outside world, and in the indigenous communities.

Yagé is not separate from the way we have lived for millennia, and the struggle we have faced for the last 500 years. It is not separate from the other plants we use for medicinal purposes. It is part of all this. We live an integral life. And all the things that have come to disrupt that, from armed conflicts to extractive businesses, to sicknesses, yagé helps us to survive and maintain our harmony.

There are laws beyond those passed at the United Nations. There are universal laws. Spiritual laws. We have rights as indigenous people. We have the right to live in our lands in a good way and protect our territories. So when we talk about yagé and our spirituality, we are talking about all of this.

We drink yagé with our grandparents to connect with all this, and to follow their instructions that have been passed down. It is not easy. But that is why we drink yagé, to connect to Mother Earth and stay on our ancestral way of life. So when we hear about corporations globalizing ayahuasca, commercializing it, we simply ask, "How?"

How are you going to do that? And is this going to actually help people? Or is it just going to continue to do that damage that that system does? Because when you use these sacred plants without spirituality, you run the risk of causing even more sickness. And that is what we see happening in the outside world right now.

There are many sick people. Every time I go to Europe, I see so many sick people. They come to us and they ask us questions. How do you use the ayahuasca? And every day there are more and more people interested in it, more people studying it, more and more "experts." But none of that is the solution. In fact, it's just causing more problems.

Why? Because they are still not taking responsibility for their actions in this world. Sacred plants like yagé can only be used under a responsibility. Without that responsibility, it's just a cup. We drink yagé to keep resisting the different strategies used to extract, to destroy, Mother Earth, and even to face this attempt to exterminate us.

This is what is it to talk about the essence of the sacred plant that is yagé.

OM: OK, thank you, Miguel. What I want to ask you about now is the psychedelic renaissance that is sweeping the modern world. What is the best way for people to engage with substances that open their minds up in this way? And what are the problems we face?

Miguel: It complex to talk about all sacred plants because we are just five communities of yagé in the Amazon forest. Throughout the Amazon there are many different sacred plants that are used, mambe, rape, San Pedro, many, many plants. And as you say, now these plants are starting to be used in the Western world.

People want to be cured, or cleaned, by these plants. But what is important for us, is that there is a lack of knowledge in the outside world about yagé and the spiritual side of life. We would like them to listen to us, to the indigenous people, because we have the traditions and experience of working with the sacred plant yagé. Listen to us, and respect what we say. And this way they will learn how to live life right.

It's not just about respecting us; it's about respecting the plant itself. Because right now if you look at the Internet you will see publicity about yagé that is selling it like a rock concert. Commercializing it. But we don't believe yagé is going to cure anyone that way. We have been hearing about things like manipulation of people, particularly of women, coming out of these situations. And this is something that worries us greatly because it just furthers the harm to society.

People are also mixing all the different plants together in ways that has never been done before. They start with one and then go to another. But they never get to where they should be going. If I want to cure myself, I have to finish the process with this plant, and not just jump to another. It shows a lack of confidence in the plant. And lack of respect.

We want to be clear with the outside world. Yagé is not the cure for the planet. The cure for the planet is the consciousness of each person. That is what is going to cure the planet, help Mother Earth, and will be the final solution to this system we are in. If we are not conscious in our actions, we are not going to see a global transformation, or even a personal curing, no matter how much yagé we consume.

The way they are doing things now will just cause more problems. We are hearing about the globalization, the medicalization of yagé. Yagé for depression. Yagé for post-traumatic stress syndrome. Yagé and other plants to cure a wide variety of things. But if these plants are in the hands of people without consciousness, all of these will just result in more sickness.

And that is because again, yagé is not the solution for all of humanity's problems. Consciousness is the solution to all of humanity's problems. Yagé is used by indigenous communities to live in harmony with Mother Earth. But we also just can't open up our territories to the whole world. And again, that's because the salvation is not with the indigenous communities either, it's within each one of us.

So why is there so much commercialization of yagé in the world right now? Because no one is telling the truth that yagé is not the solution for everything. It is being manipulated to make money. They are saying come here and pay to drink yagé and you will be cured, you will be healed. It's a lie.

There is also this huge interest of different companies and corporations that want to put ayahuasca in a laboratory and study it. They want to discover what's in it so that later they can sell it in pharmacies and say, "This is the cure." But that's not going to cure anything.

It might calm somebody, just like other pharmaceutical medications that they sell in pharmacies. It's like an aspirin. They want to convert it into this kind of product. That does not actually cure, but just calms the pain. And in this way they are scamming people.

I will not tell you that you should bring all the people here so that we can cure them. Nor that I will go there and cure everyone. Yagé is not the cure for everyone and everything. I want to make sure that yagé stays here in the territories, planted and used by us. The way things are going we will be erased and the yagé will only be found in pharmacies. So we have to make sure this does not happen.

The system would love to replace the garden, the forest, Mother Nature, with a pharmacy. And it's even happening here that some people think that cures come from the pharmacy. But we know that the pharmacy just brings more sickness. We know that in the garden, in the forest, in Mother Nature, in the plants, is where the real cures are.

What we worry about the most, what we see in the outside world, is that the yagé will no longer be used in the maloka. It will be put in a capsule and sold in the pharmacy. It seems so much easier right? But there is no cure there. What we want is to keep guarding our territories, that is where the real treasures of life are, that is the real cure. And I'm not just talking about the material, I'm talking about the immaterial. What you see, what you perceive, what you feel when you drink yagé, is that connection that should not be broken.

OM: OK, very good. It's a very deep subject but I just have one more question for you. If yagé is not the cure, mushrooms are not the cure, but the consciousness is the cure, can you tell me a little bit more about what kind of consciousness this is and how we get there?

Miguel: For us watching everything that is happening, and not just with yagé and medicinal plants, but what is happening around the planet, we see that we have a system of manipulation. I'm talking about capitalism. I'm talking about globalization. And here we are watching it and resisting it in our own territories.

Not too long ago I was invited to speak in a conference, an event about ayahuasca, and there was all this talk about more investment in the Amazon. More development. But what kind of development? When they are talking about investing, they are not really talking about development, they are talking about making money. And after all the development that has already happened in the Amazon, all it has done is hurt the forest and the indigenous communities. Real development would be to make reparations. But that is not what they are talking about.

So, when I talk about the real cure, I'm talking about a change of consciousness. Right now we are living in a world that is going through changes. When I drink yagé and see things in a more spiritual light, I see these changes happening. How is the world changing? Right now we are in an era of consumerism. But the problem is the more we consume the more we contaminate.

I see all these people going to conferences on how to stop climate change, how to stop industrial damage to the planet. But despite all these conferences, nothing changes. Every day the environmental crisis gets worse. We are running out of fresh water. We have polluted everything.

And that's because even though we are in a crisis, we keep consuming and consuming. And that means contaminating and contaminating. The more that humanity creates the more it destroys. And that is because humanity is out of harmony with the earth and is disconnected with the spiritual world.

Everyone has their territory. Everyone had their own culture originally. But something came and damaged this. The reality of being a human being became distorted. We began to only create damage to Mother Earth.

As indigenous communities, we are trying to show through example how to take care of Mother Earth. But the people are still not listening. They still see us as undeveloped, as poor. They see us as people that do not want a better life. But what exactly is this better life? What do they want us to be looking for?

If we have clean water, if we have healthy forests, if we are producing food in our territories, we are living well. There is no need for anything else. No need for petroleum, gold, or carbon. If we did need those things, we would lose the harmony with our territory. We would have to break the connection we have with Mother Earth. Because for us, everything in our territory is sacred. All of it is alive, it is full of spirits, and we must protect it.

So, I am saying to you that in the indigenous territories we know how to live well with Mother Nature. We don't need these consumer items, this luxury, of the outside world. Apartments, cars, industrial products, all of that. We know that it all creates damage. And so I want to say that at the planetary level, not just those out there in the modern world, but all of us, because we are in this system also and we have indigenous people living in different states, different levels, of this colonization. We all have to make this conscious transformation withing each of us.

What is this transformation? The cause of much of our problems is that within us, we look for superiority. I want to be better than my fellow human. I want to be superior than my brothers and sisters. I have cultivated my mind to be individualistic, to be self-centered, to be business oriented, to think just in business deals. All of this creates a disconnect with Mother Earth. And this is the root of our problems. This is the part that we have to change.

If we can make that change. If we can reharmonize with Mother Earth, the world will transform.

3

Shamanism and Spiritual Evolution

Jungle Svonni, Sami Shaman

Jungle Svonni is a Sami Shaman from one of the most isolated reindeer herding districts in northern Sápmi, about 200 kilometers north of the Artic Circle. The Sami are the only recognized indigenous group in Western Europe. In his mid-20s, Jungle traveled to the Amazon to study plant medicine and ended up staying for over seven years. His intent was to recover knowledge that was lost or on the verge of extinction among his own people. Now back in his homeland, Jungle speaks and teaches at a variety of conferences and centers.

∞

My first memory in life was that I was going to become a shaman. This was before I could speak. Later, when learning to speak, I learned that the word for what I was to become was *"Noaidi"* (shaman in my language, Sami).

But my story began long before this. Let's start at about 500 years ago, when my country Sápmi was slowly starting to be invaded by Swedes, Finns and Norwegians, and by Russia, which was a lot farther east. First came soldiers, tax collectors, accompanied by Christian missionaries. The foreign invasion resulted in our religion/spiritual beliefs being made illegal by the invaders. That's how to conquer a land.

Practicing shamanism or anything related to it was punished severely, sometimes with death. Things related to shamanism were things like having a drum in your possession, for example, or to *yoik*, Sami traditional singing. So, shamanism went into hiding. When something goes into hiding, it doesn't get passed on.

What makes an indigenous people indigenous is their spiritual connection to the land they live on. And that is what shamanism is.

The difference between a religion and shamanism is that shamanism has no dogma, telling you how to live. There are no commandments from God or gods. Shamanism is the spiritual understanding of life and the universe through nature. It is the understanding of and the connection to the divine life force in every tree, in every plant, in every lake, rock, river, sea, animal and human. It is also understanding how to work with these primal forces to heal yourself and others, to make life better. To make magic.

In my country, Sápmi, medicinal plants were used like everywhere else on this planet. A few psychedelic substances like mushrooms, both amanita and psilocybin. But a shamanic trance was mainly induced by drumming and *yoik*, and also meditation.

Although reindeer herding is relatively new to us—400 to 500 years—reindeer have always been the most important thing to us. Before the reindeer herding, we followed and hunted wild reindeer. Our colorful clothing is something very shamanic. Since we are deciding about whether a reindeer will live or die, we dress this way to make their lives beautiful. Of course, a lot of Samis today have no idea about this.

In almost every language, the most common theme in the songs is love, but in our *yoiks*, the most common theme is reindeer because without them, there is no life and no love.

Now back to modern times. When I was born in the late 70s, very little of our shamanism remained, only fragments. My grandfather was one who "knew," but he was not known for it. It was kept more like a secret because of centuries of persecution. And the ones before him in my lineage also had this knowledge. My father was the one who broke the tradition, but it returned in me. Since my grandfather died when I was 12 years old, there was no one left to teach me. Only fragmented knowledge remained scattered all over Sápmi.

This spiritual connection to the land, shamanism, is like a common root that holds an indigenous people together. This root has to be thick and strong. Growing up, I saw this root become dangerously thin and weak like a thread. How people in my tribe more and more started to think like the colonizers. Our rate of suicide is very high as a consequence of this loss of identity: our spiritual connection to the land we live on, to the reindeer and shamanism. Seeing all this as a kid made me so sad that I didn't want to live anymore, but miraculously I lived and grew up.

In my mid-twenties, I took a radical decision. The last place on earth where I had any desire to go was the Amazon, but I realized that that was exactly where I had to go because the Amazon was too big and dense for the missionaries to penetrate and eradicate shamanism. So, I had to go from my village in Sápmi, roughly 200 kilometers north of the Arctic Circle, to the Amazon to learn my own culture! Luckily, I did not know how long I would end up staying there, before I would feel ready to return. I ended up staying there for about seven years, most of the time in Peru in the Iquitos area. At the beginning, I also spent some time in Ecuador. In the Amazon, I met with many shamans and studied extensively with one of them. It was there that I experienced a plant medicine practice commonly called "dieting" in the Amazon.

I also took part in hundreds of Ayahuasca ceremonies. After some years, I led hundreds of Ayahuasca ceremonies myself. San Pedro (a cactus) was another plant that I used frequently.

Now, six years after returning home, there is a question that, to my great surprise, I have been asked many times. How is the shamanism different between Sápmi and the Amazon? That question never fails to surprise me. The answer is that there is no difference. No matter where you stand on this planet, the natural laws are the same. The divine force within everything is the same.

Since I know where I am from, my heart belongs to the Arctic, Sápmi in particular. I did not conduct any ethnological studies of the tribes in the Amazon. I did not learn any of their *icaros* (shamanic songs) in native languages or in Spanish. None of that held any interest for me. Instead, I connected to the source: the divine force of nature. So naturally, my *icaros* appeared as Sami *yoiks*. Through the ayahuasca and other plants, I connected to my own ancestry and the mountains, rivers, fjords, lakes and forests of Sápmi. Many people who didn't even know I was Sami could, to their surprise, see reindeer, moose, eagles and snow-covered mountains in their visions during my ceremonies.

Finally came the day I had been longing for so many years. I had a vision telling me that I could go home to the mountains, northern lights and midnight sun. That was a very happy day.

Upon arriving in Sweden, on my way home, almost in Sápmi, because of a package of San Pedro, one kilo of dried and powdered San Pedro to be precise, I was jailed and portrayed in the media as a member of the international mafia. In my cell, I was thinking about how little things had changed in the last few hundred years. My ancestors were persecuted and jailed for communicating with the divine force of nature just like me. My ancestors were persecuted and jailed for using a tool in their work and got their tool, their drum, confiscated just like I had one of my tools confiscated, the San Pedro.

Not much progress in human evolution.

However, after appealing to the Royal Court, the second of three levels in the Swedish court system, I was released from jail after 18 days, but the charges against me were not dismissed. So, the legal case continued for one and a half years until I was found innocent due to a decision of the Swedish Supreme Court. One important reason why I won this case was the massive international support I received, something that strongly affected the energy of this case. Ceremonies were held in Peru and Canada, amongst other places, for my liberation. So, I became the first Sami shaman in history to win over the government.

Now looking to the future. What does the future hold for shamanism, for Sápmi and for this planet in general? Well, those things are closely linked together. Without shamanism, the world has no future. We've been at the critical point for a long time. Our natural resources — ocean, air and land — are getting more and more depleted and polluted, and some people are making profit from war and violence.

During the Middle Ages, Western society put its faith in religion. Then, later on in history, people put their faith in democracy. Now, in the most recent years, I see a new trend. More and more people have started to put their faith in their personal, spiritual evolution. Yoga, meditation, shamanism and other spiritual practices are increasingly popular. This, I am convinced, is our only way forward. Without a heightened spiritual awareness and a stronger spiritual connection to nature on a global level, we don't have any future at all.

And what can be done by society, by us, to achieve this? To start with, take away medieval laws that are restricting people in their shamanic/spiritual practices. These laws are greatly endangering our existence as a species on this planet. Secondly, I think allowing more space for spiritual, nonreligious practices in the public education system would be a very good idea. Also, the only sort of government I can imagine in the future would be one where spiritually aware officials sought spiritual counsel before making decisions.

Our spiritual connection, shamanism and our level of spiritual enlightenment are all intimately related and they must be our number one priority, if we are to survive as a species.

4

Psychedelic Feminism: We Are Wildlife

Zoe Helene of Cosmic Sister

Zoe Helene, MFA, advocates for women, wildlife and wilderness, and the right to journey with sacred psychoactive and psychedelic plants and fungi—our coevolutionary allies. She founded Cosmic Sister, a feminist collective and creative studio with interconnected advocacy projects including Ancestor Medicine, Temeno GAIA, and Psychedelic Feminism, which she originated to support Women of the Psychedelic Renaissance.

Zoe believes creating a true balance of power across the gender spectrum—globally—is the only way humans (and nonhumans) will survive and that it is our moral responsibility to protect and defend the rights of all earthlings and Earth herself. She is a spiritual agnostic who identifies as Indigenous in diaspora in honor of ancestors who developed sophisticated, nature-based, female-led, entheogen-elevated mystery traditions dating to the Neolithic era or farther, and is proudly part of the global herstory, reclamation, revival, and rematriation movement. Zoe grew up exploring primary rainforest and teeming tidepools in Aotearoa (New Zealand) and was mentored by the legendary theatrical costume designer Patricia Zipprodt. An artist and educator, her work has been covered by top-tier media outlets. She is happily married to ethnobotanist Chris Kilham.

∞

Humans can be exceptionally creative, clever, and courageous. We're blessed with an insatiable curiosity and a keen desire to communicate—a *superb* combination—and our propensity for love is profound. We have immeasurable potential and truly are capable of greatness. And as a species, we have lost our way.

Our species suffers from megalomaniacal delusions of grandeur. The constructs we have created of human supremacy and male supremacy have proven deadly—insidious, yet omnipresent—and we are collectively enslaved by these forces as if enchanted.

Human supremacy is one of the most destructive (and self-destructive) constructs our species has manifested. The moment we proclaimed ourselves better than, or more important than, other species was the moment we started down the fast track to devolution. Other cruel forms of supremacy such as sexism, racism, classism, ageism, and ableism are also constructs, cut from the same "othering" cloth, and mutually dependent. They're predatory, ultimately cannibalistic, and they brought us to where we are today—The Age of Extinction.

Male supremacy is so deeply entangled with human supremacy that it is difficult to distinguish one from the other. Patriarchy is a euphemism for male supremacy, and while no one alive today is responsible for creating patriarchy, everyone and everything on Earth is hurt by it. If we continue living under the delusions of these constructs, we destroy ourselves and all of life on earth as we know it.

Not everyone is violent, power-hungry, grotesquely greedy, and asleep at the wheel—but too many are. The reality is, many people who have risen to power appear willing and able to destroy this genius of a planet we inhabit, spending precious resources to develop more ways to wage war, and explore other moons and planets so we can colonize, overpopulate, and destroy them, too.

The Fates are not amused.

For those of us who know better, *this hurts*. We do our best to keep up with the chaos and maintain hope, to propose, develop, and support solutions, and to spread the word—even as our guts (and mounting evidence) tell us that we're fighting a losing battle. We're exhausted, but we *know* it's time for radical positive change.

Witnessing the Age of Extinction cuts to the quick. Some people choose to handle the onslaught of grief (for what has been lost, may be lost, will be lost if we continue down this path) by promoting the "if we destroy life on earth as we know it, life on earth will continue" campaign, which feels like a cop-out to me. I understand it may be a scientifically plausible scenario in a romanticized dystopian post-apocalyptic kind of way and that detached defeatism is fashionable in some circles, but shirking accountability isn't hip or cool.

Also—it isn't all about us.

How about we just don't destroy life on earth as we know it? How about we just—don't be that species?

We need to face our role in this extinction crisis head-on, and we need to do it now. It's time for our species to rapidly evolve (culturally—our behavior is the issue) so we can be better citizens of the planet we all depend on to survive.

The good news is, constructs don't hold up to psychedelic scrutiny, and we have these brilliant, mysterious co-evolutionary allies—the sacred psychoactive and psychedelic plants and fungi—reaching out, checking in, calling us to task. These medicines—our kin and our teachers—are truth-serums. They can help us spot the con and, if we're willing, navigate the opportune moment, as they have for time immemorial.

My Ancestors Were Tripping

My ancestors are aboriginal people of the Hellenic Republic, known as Greece to much of the world, known as Hellas to those in the know. Many of us are in *diaspora* (a word from my ancestry that, roughly translated, means "to scatter across" or "to disperse") because of life-threatening circumstances and the ravages of a series of savage wars in succession. We did not want to leave.

We come from the ancient tribe of Arcadia in highlands at the center of the Peloponnese, although our tribal boundaries shifted in later years when Karyes (our home village) was conquered and annexed by the Spartans. Today Karyes is in the state of Laconia, and Sparta is the capital. We bridge two tribes. Karyes is only 20 miles from the ancient city of Mycenae, and in the old days, our dialect was Mycenaean. Mycenae translates to "the mushroom people." (Yes! The word mycelium comes from the Greek word *mykes*, which is fungi.)

Most people know Mycenaeans because they won the Trojan War, but more importantly, they are credited for inventing the first alphabetical writing system, a revolutionary syllabic script known as Linear B, that propelled the period from Prehistory (before written language) to Protohistory (the invention of written records in a culture). With the invention of an alphabet, Myceneans could record stories, reports and other key information that we are able to decipher today. Through the ages, this language evolved into modern Greek.

But Mycenaeans didn't invent the first alphabetical system in a vacuum, and it didn't happen overnight. Mycenaeans and Minoans are culturally close and genetically close to identical, and the Mycenaean alphabet (Linear B) evolved from a Minoan predecessor (Linear A). Also, it seems highly likely that magic mushrooms empowered this evolutionary leap, so I see this as a shared credit.

What distinguishes Mycenaean DNA from Minoan DNA is a Scythian and Siberian ad-mixture (genetic marker). Siberia is where the word "shaman" originates, and many people associate the iconic white spotted red psychoactive mushroom Amanita muscaria, which is native to the Peloponnese, with Siberian shamanism. In both Siberian and the Peloponnesian culture, deer and wolf-dogs pulled sleds or chariots. They would most certainly have shared knowledge, theories, tools, techniques, including the arts and craft of shamanism, as well.

Myceneans took their rituals very seriously, and Amanita muscaria was not the only psychoactive mushroom available. Psilocybe cyanescens, the potent psilocybin mushroom commonly known as the liberty cap, is also native to the region. Myceneans, who were talented artists, architects, and engineers, built elegant *tholoi* (large, vaulted stone ceremonial spaces with refined acoustics) in the precise shape of a liberty cap, which was also the shape of their warriors' helmets.

The Myceneans were also keen travelers, and thanks to strong, streamlined ships and advanced navigation techniques, their trade routes were expansive. They readily engaged in talent-sharing in all directions and were enthusiastic about all manner of medicine hunting. Sacred psychoactives were very much a part of Mycenean and Minoan culture, as shown on countless works of art—especially gold "signet" rings—depicting, in excellent detail, scenes of medicine women (and sometimes men) dancing ecstatically with sacred plants and more. The Myceneans were tripping.

But my ancestors were thriving in the region long before the Mycenean era. Evidence dates back to flourishing communities during the Neolithic (New Stone Age), farther back to the Mesolithic (Middle Stone Age), and farther still to the Paleolithic (Old Stone Age). In 1960, the Petralona skull, a human cranium that dates back to 350,000 and 200,000 years ago, was discovered in a cave in northeastern Greece. In 2002, footprint fossils that are (potentially) 6.05 million years old were discovered in Crete, which would have been attached to mainland Greece at that time. More likely than not, sacred psychoactives were also present in these earlier cultures, too.

Arcadian Artemis

Arcadians are considered one of the oldest Greek tribes. Geographically, Arcadia is somewhat of a cultural refuge, set high on the Parnon (or Parnonas or Malevos) mountain range in the Peloponnese, which is often called a peninsula but technically is an island — connected to the mainland in two spots.

It's no wonder early humans migrated to the region. There are at least 8500 caves in Greece, which provided shelter from harsh weather, predators, enemies, and invaders, and served as homes, community centers, and safe, beautiful ceremonial spaces. And in the Arcadian region, a network of calcium-rich freshwater underground springs runs in and out of the limestone mountains — calcium carbonite, which is the primary component in limestone, supports life in too many ways to mention.

Gaia's inspired abundance is manifest, with dense forest and rolling, pastoral hills. Miles and miles of diverse, old-growth deciduous forests bless the mountaintop plateau, ideal for forest foraging, and later, forest farming. Bountiful oak, chestnut, walnut, black pine, wild olive, laurel, cypress, heather — and many, many more — have medicinal and/or food value. Nut trees in particular provide a rich source of healthy oils, protein, and fiber — a generous life force for early people during winter months.

Drys (oak trees) were supremely sacred, and the first *Dryads* (a spirit being in human form that is irrevocably bound to the life force of a tree) were spirits of the oak. This opened the ancients' psyches to the spirit in all trees, then all plants, and the word Dryad expanded to embrace them all. The life force of a Dryad is bound to the life force of a tree or plant. Dryads rise out of the soil fully formed, and their plant sprouts from the spot of earth where they were born. When a Dryad's plant dies, they die, too.

The name of our home village, Karyes, means "walnut," and my family name "chestnut." The ancients say that acorns and beechnuts, or "mast," were the first foods eaten by humans.

Long before the Olympians, our primary goddess was the Indigenous Aegean Carya, Lady of the Nut-Tree. Later, she merged with Olympian Artemis and together they came to be known as Artemis Karyiastis (Artemis of the Nut Tree) or Arcadian Artemis. Arcadian Artemis is a complex combination of both goddesses who, among other things, embodies *eleftheria* (freedom and liberty). Her realm is where the wild things are in the "real" world wilderness, the otherworldly wilderness within, and where these worlds coexist—which is everywhere and is open to exploration. In these realms, she is Lady of the Beasts, deer goddess, bear goddess, daughter of the moon wolf goddess, protector and defender of wild beings and essential freedoms.

The Caryatids, which means "maidens from Karyes," were *caryatidai* (priestesses) dedicated to Arcadian Artemis. These young women were ritualists known for making music with exquisite harmonies and irrespirable complex rhythms, and for dancing down the moon and "turning into plants"—becoming spirits of sacred flora and fauna. Their mentors, older medicine women who were wisdom-keepers and masters of the craft, and who had been Caryatids in younger days, sometimes joined in ceremonial dances for Arcadian Artemis. The dances, which were considered difficult, are unique to the village. These moon priestesses, my ancestral sisters, are immortalized in the masterpiece sculptures at the Erechtheion temple on the Acropolis of Athens as the six Caryatids—and they once held phiale medicine vessels decorated with bees, acorns, and beechnuts, honoring Arcadian Artemis in all her wild "belonging to no one" forms.

In my tribal homeland, the veils between the worlds are fluid, flowing and shifting between journeying realms such as Artemis's wilderness, Persephone's Underworld, Ariadne's labyrinth, or Hecate's time magic. Athena's owl keeps watch. Hestia, goddess of "home" as soulfire, plays a vital centering role.

Women made and served sacred medicines and led ceremonies and rituals in honor of shape-shifting goddesses and other divine beings who championed seekers on mystery journeys. Deities could become sacred flora and fauna, and sacred earth elements and phenomena too, as could untold wild mythical beings in their mystical realms.

Our tribe also had the Arcadian Mysteria, which was very much like the psychedelic Eleusinian Mysteries. The primary goddess of the Arcadian Mysteria was Despoina, a free-flow epithet for several goddesses — especially Persephone, Demeter, Aphrodite, and Hecate and Arcadian Artemis (meaning she was or could be all or any of them).

I am proud to be part of a global movement working to discover, reclaim, revive, and rematriate (liberate!) mysteries that thousands of years of patriarchal reign couldn't disappear.

I was born a freedom fighter and will die a freedom fighter. The Greece national motto is *Eleftheria i Thanatos* (freedom or death), which became the cry of the Greek revolution against the Ottoman Empire. On my mother's side, we were taught about how the Greeks kept the language alive on pain of death during Ottoman rule and how village women trapped by Ottoman troops chose to sing and dance off a cliff (with their children and infants), rather than be captured. For more than 400 years, slavery was one of many atrocities committed by the Turks on the Greeks, and tens of thousands of females were sold to the harem sex slave trade (there are records of sex slaves sold as late as 1908).

I was also raised to be fully aware that we are animals, and this is one of the primary lenses through which I experience myself and the vast living universe that I play an infinitesimal part in. I have come to realize we are also *wild* animals. People may be conditioned or even brainwashed to believe we are "domesticated," but domestication is another human construct, and like all constructs, it has a shelf-life. We may be under a social spell of relative tameness, but wild things must be free, and we have what it takes to lift the enchantment if we want to.

I learned early on that wild species do not thrive in captivity and that most wild species will go insane and/or die if they are caged for long periods of time. As a girl, I was fully aware that I was a wild being, and I was fully aware of the "wilderness within" at all times. My inner world was especially vivid and creatively prolific when I was exploring wild outdoor spaces (which I did a lot) or in artistic flow-state (most waking moments). Both inner and outer realms were present and accessible to me. There was no separation to bridge. I am still that person.

When we say "humans and animals," or talk about "human rights" and "animal rights," we perpetuate a false separation that enables human supremacy to spread and flourish. Once we *know* that we are animals—wild animals, wildlife, earthlings—perhaps we can begin to learn who we are. And perhaps from that perspective, we can learn how to coexist with other earth species—to grow with each other, share the abundance, harmonize rather than monopolize.

We humans can be very "in our heads." We tend to categorize and compartmentalize and "other," which can be helpful in many ways but can also foster delusion. Even the term "environment" is detached language, as if we are somehow separate from the environment or the environment is a setting for us to play in or a set to play upon, when in reality we're all part of the environment, actively participating in an interconnected ecological entity.

We are part of Earth's ecosystem, as Earth is part of the ecosystem of our solar system, our galaxy, our universe, and beyond. We are not separate from or different than nature. We are not superior to or more important than other lifeforms, and we do not have the right to rule over them. These longstanding, far-reaching constructs have caused great harm and led us to where we are today—the Age of Extinction.

Destructive and ultimately self-destructive systematic social constructs are symbiotic—it's impossible to cure one without curing the other because they are co-evolutionary. Systematic social constructs die hard. Once deeply indoctrinated (by choice, coercion, or force), psyches often assimilate the poisonous concepts as if spellbound, as if they always were and always will be, as if there could be no other way.

Ideas that are widely accepted may have a life of their own— but they do not own us. Ideas that aren't helpful or healthy—for ourselves, other humans, other species, and Earth herself—are just that, *ideas*. Sacred psychoactive medicines can help us spot the con in every drop of toxic indoctrination and conditioning by exposing their character, origin, structure, catastrophic trajectory, and Achilles' heel—sometimes in screaming technicolor.

Each and every one of us is responsible for breaking spells that ensnare our psyches, for ridding ourselves of hardcore constructs and systems that hold us back and down, and for setting ourselves free. Each individual is responsible for their own self-liberation and healing because, while others can guide and support, the work is ultimately our own. Once we have fully integrated the sacred death and rebirth, we are also responsible for spreading the word and contributing in what ways we can to the liberation and healing of our species.

Our goal must be to treat all of Earth as a living being and a sacred space. The top priority—above all others—must be Earth rights.

Temeno GAIA

Temeno GAIA is a concept that emerged, fully formed, from my ancestral memory.

A lifetime of service and devotion to the arts, women's rights, and the rights of wildlife and wilderness led me to Temeno GAIA. Decades devoted to labyrinthine ancestral research—especially the Language of Psyche as universal language. Decades of deep conversation with cherished elders and trusted family and friends who are also on their Ancestor Medicine journeys—many of whom are also Indigenous cultural leaders. Decades of intentional journeying with sacred psychoactive medicines (including cannabis, magic mushrooms, ayahuasca, peyote) and in community with others on the Psychedelic Feminism path. Plus, some 16 years of life-changing frontline ethnobotanical field experiences with my sweet husband—sharing time with Indigenous peoples in vast deserts, primary rainforests, high-altitude mountain ranges, and remote islands. This never-ending, ever-hopeful life work is equal parts heart-healing and heartbreaking, and I have nothing but gratitude for the sacred cycles of death and rebirth that have brought me to the present moment.

The catalyst for my "initiation" into Temeno GAIA was falling in love with a pair of wild silver foxes, who brought their adorable kits to visit one moonlit midsummer's eve, in juxtaposition with the knowledge that my neighbor would—and legally could—shoot and kill these magical beings if they made the mistake of crossing that arbitrary human territorial delineation we call "the property line." And I thought, *What good does it do to designate regions on earth as "protected sanctuary" when the nature of nature is freedom and expansion?*

My ancestors felt no separation from nature. All living beings—fantastical and otherwise—had spirit, as did objects, places, earth elements, and earth phenomenon. I have come to embrace these energies and entities as *myriad kin*, and this worldview as treasured cultural heritage.

The knowingness that *we are nature*, that *we are wildlife*, and that all life on Earth and Earth herself is sacred space is Temeno GAIA consciousness.

Etymology, or the study of the origins of words, is a go-to clue-finder in my personal Ancestor Medicine. Language is born of a desire to communicate what matters. When a foreign culture "borrows" a word or phrase from an originating culture, that tells us something about the borrowing culture. Also, important dimensions of words and phrases are often lost in translation, in part because they are taken out of context, in part because the borrowing culture is not of the originator's world. When cultures that borrow words and phrases (and symbols, myths, philosophies, etc.) are a dominant power of the times, their skewed and/or reduced interpretations are reinforced by stakeholders, spread widely, and become the standard, with deeper dimensions overshadowed or buried—until the originating culture is strong enough to discover, reclaim, revive, celebrate, *rematriate*.

My people are ready, and this is happening. Approximately 30% of the words in modern English have their roots in the ancient Greek language. The words *psyche, temeno,* and *Gaia* are among those words.

"Psyche" is *that which animates our bodies*. Psyche does not mean "mind" (as in, "mind manifesting") or "soul" or "spirit." Psyche is *all that we are* aside from our physical, sensual self. All living beings have psyche, as do earth elements, earth phenomena, Earth herself—and each divine being in the Psychedelic Pantheon. Psyche is also a goddess who was born a mortal. She's a divine feminine personification of "that which animates our bodies," or "breath of life."

Over thousands of years, my ancestors created the Language of Psyche, a rich, fluid, psychedelic-informed and inspired poetic universal language for exploring and expressing the human condition. This language resonates with seekers from a wide range of genetic ancestry and cultural heritage even today.

Sometimes essential life forces show up in visionary states as personifications—anthropomorphic or otherwise. For many, they present as divine feminine, divine male, divine non-binary, or gender-fluid archetypes. These personifications empower the storytelling arts, and storytelling arts are innately and marvelously human, relating to each other as "characters" in relationship to self, others, Earth, and life itself. Archetypes aren't rigid—they are more like musical notes or a box of paints to create with and express through.

Occasionally, divine beings from my ancestry visit me in ceremony because they're part of who I am, but people who aren't familiar with our particular poetic language might experience archetypes and symbols from other tribal traditions—as these languages express human condition, and we are human.

"Temeno" means sacred space. It can be human-made (a tipi, maloka, nakamal, temple, yurt, treehouse, or any human-made enclosed space) or it can be a sanctuary outdoors. Temeno is the original, indigenous word in Minoan and Mycenaean Greek and the earliest "attested" form of the word as of this writing.

A temeno is often a setting outdoors that is special in some way. For instance, a plateau on the edge of a mountaintop with a long-range view, a sun-kissed clearing in an old growth forest, a waterfall where two rivers converge, a crystal spring, a womb-like cave, a field of poppies. Your garden can be a temeno. Your secret swimming hole can be a temeno. Wherever the space, a temeno must be respected, protected, and life-affirming, so that seekers can feel safe to surrender to (sometimes extreme) vulnerability.

"Gaia," as primordial goddess, is a fundamental force of the universe, rarely (but sometimes) depicted as female in human form. She is the mother of all Life, she is Wild Living Earth in the Wild Living Cosmos. A conscious, intelligent, collaborative, ever-evolving divine creative work *born of Love*, in which all earthlings and earth elements and phenomena are active participants. Gaia is central to my matrilineal creation myth. She is in my DNA, and I proudly celebrate her as our first Divine Feminine.

Psyche, temeno, and Gaia were born of a culture that understood how sacred psychoactive and psychedelic medicines from the earth could help us experience a shift of perspective that *was essential to the human experience—to make us better humans*. These words of power—and many more—are the work of Indigenous ancestral wisdom-keepers. If we journey with the intention of living in Temeno GAIA consciousness—even if we know we can never do that perfectly, even if we know that, sometimes, matrix realities (an intentional oxymoron) demand compromises—then maybe, just maybe, we have half a chance.

The following words help me call Temeno GAIA consciousness into my medicine experience.

I am nature.
I am nothing but nature.
I am an animal—a wild animal.
I am wildlife, Indigenous to earth.

You are nature.
You are nothing but nature.
You are an animal—a wild animal.
You are wildlife, Indigenous to earth.

We are nature.

We are nothing but nature.

We are animals — wild animals. Wildlife, Indigenous to earth.

Our species is 8 billion strong and still breeding exponentially.

Our species is one of some 8.7 billion unique species alive on earth today sharing this precious planet.

We are no better than or more important than these other species and we do not have the right to own, dominate, control, enslave, exploit nonhuman earthlings, or drive them to extinction.

We are no more important than essential habitats and earth elements and we do not have the right to own, dominate, control, exploit or destroy them.

How fortunate we are to be alive! How fortunate we are to be part of this fragile, vibrant, conscious spinning miracle of life. How fortune we are to contribute to the symphony of symbiotic diversity that is life on earth. How fortune, how fortunate, how fortunate we are.

It is our responsibility to heal, self-liberate, learn and educate. It is our responsibility to be in right relationship with Earth and all earthling species.

In joy and with wonder, hope and love.

Know Thyself Seekers

Know Thyself — that jewel of Indigenous psychedelic wisdom — remains as precious today as it was thousands of years ago.

"Know Thyself" is carved above the entrance to the Temple of Apollo at Delphi, where people made pilgrimages to consult the Oracle of Delphi, a great ancestral healer dating to 8th century BC, who was also known as the *Pythia*. Some believed she channeled prophecies directly from sun god Apollo, the All-seeing Eye, when he possessed her, but I prefer to see her as a strong, talented medicine woman in a highly influential and prestigious position. Apollo is Zeus's favorite son, and Zeus is the King of the Olympian Gods.

But Delphi, which the ancients named "The Womb of The World," existed long before Olympian Apollo was even born. A Mycenaean settlement existed within the sanctuary area around 1400 BC, and before that the region was a sanctuary devoted to the Gaia, mother of all Life, built around Castalia, her sacred spring at the center and birthplace of the world, and marked by a sacred stone called the *Omphalos* that represented Gaia's cosmological bellybutton and the center of the world.

Some 9000 years ago, during the Neolithic period (6800 to 3200 BC), ceremonies and rituals that included prophecy were held in Gaia's honor in the Corycian Cave, a few miles up Mount Parnassus from where the ruins of the Delphi are today. The cave has a large, womb-like room with a perfect plateau for gathering around a fire pit, and journeyers brought decorated "disposable" ceramic drinking vessels into the cave—perhaps to hold a dose of sacred medicine. Ceremonies continued to grow throughout the entire Mycenean period (1600–1100 BC) and beyond.

In the eighth century Apollo slew Gaia's son, a giant sacred seer-serpent-dragon named Python who guarded the center of the world, and he took the Gaia's sanctuary for himself. Others say Apollo slew a sacred seer-serpent-dragoness named Drakaina, but either way, this signaled a significant power shift in the region, and a new temple dedicated to Apollo was built on top of where Gaia birthed the universe. This happened to many divine feminine strongholds around the world—first with harbingers of patriarchy sharing space but taking the lead, then agents and enforcers sent by patriarchal monotheistic religions (plural), who asserted dominance more forcefully.

Even so, Gaia remained a formidable presence at Delphi, as no one dared to disappear her. (Today, an estimated 700,000 people each year visit Delphi, where they can witness the Gaia's Omphalos stone at the Delphi Archaeological Museum.) "Know Thyself" seekers came from far and wide, showering the Pythia with lavish gifts to win her favor, and over time, Delphi amassed great power, wealth, influence—and an extensive art collection—rising to its height around seventh century BC.

The Pythia, a witch, priestess and medicine woman, was expert in both divination and necromancy—so she could bridge the worlds, see the future and talk to the dead. It isn't a stretch to call her a "shaman."

After both parties were purified in Gaia's sacred spring, she received individual "know thyself" seekers in the *adyton*, her private sanctuary within the sanctuary (adyton literally means "do not enter"), in "a cavern hollowed down in the depths" underneath the Temple of Apollo.

Upon entering the small, enclosed space, seekers found themselves face to face with the Oracle, who was seated on a covered "trifold cauldron" (a large, thick pot for making medicines) in an exalted position, surrounded by scented steam and gaseous fumes. She counseled in flow-state, elevated by a medley of mood-modifiers. A brew of various medicinal plants boiled and bubbled in the cauldron, which was vented, and attached to a tall tripod chair so she could sit over it without getting burned. The tripod was perched to span a deep chasm where psychoactive gaseous fumes came up from Gaia's womb through the cracks. She held a *phiale* (ceremonial medicine vessel) in one hand and a branch or several branches of leafy medicinal plants in the other, which she shook to create *rhythmos* (trance rhythms).

The Greeks were master medicine makers who understood the value of altered states, and traders with extensive expedition routes brought home exciting new plant medicines from foreign lands. The Pythia had access to a diverse collection of natural psychoactives—both native and introduced—and, like medicine women in modern times, she was expert in creating artisan tinctures, teas, ointments, and combinatory brews for the moment at hand. She also mentored others to keep the art and crafts alive and to pay it forward, in sacred reciprocity, honoring her own mentors.

The Oracle, and eventually the seeker, too, was enveloped in misty truth-serum vapors and euphoric methane and ethylene gaseous fumes rising from the grinding of fathomless faultline fractures below, trancing out to rhythms and focusing on his or her intention. Theatrics were also part of the magic. The Pythia would have been naturally charismatic, with an otherworldly stage presence. As with any medicine experience, preparation, intention, great expectations, and intense anticipation contributed to the charge.

At its height, the Oracle at Delphi grew so popular that several women took turns in succession being the Pythia. Records show that these women, who were over 50 (peri-menopausal, menopausal, and post-menopausal), offered private guidance to people in high-ranking positions, on topics ranging from love to war to peace to politics. They provided guidance on processing complex life circumstances, helped people weigh the pros and cons of critical decisions, and helped people grieve for loved ones or seek the counsel of psyches in Underworld.

Imagine being a woman going through "the change," telling powerful people the truth as you saw it, as it came to you—*truths that were not always welcomed*—while you and the seeker were both in altered states? Imagine telling a king that his obsession with possessing a married woman, for example, would lead to great war, or that his blind loyalty or obscene lust for money or insatiable addiction or mad ambition would lead to the downfall of self, family, kingdom, empire? Imagine being in the position of *guiding these people to seek and find and face their own truths?*

The only surviving depiction of the Oracle at work is a red-figured *kylix* (ceramic painting) from around 440 BCE, showing a graceful, poised, barefooted Themis, the Titan prophetic goddess of natural order, wisdom, and good counsel, consulting for Aegeus, the mythical king of Athens, on his "know thyself" journey. She is seated on a tripod cauldron and holds a sprig of sacred plants in one hand and a *phiale* medicine libation vessel in the other. Themis is one of Gaia's first children and the mother of the Moirai (The Fates), and Aegeus is the father of Theseus, the hero who followed the goddess Ariadne's Thread of Life out of the labyrinth after slaying the Minotaur, the monster within—another example of the Psychedelic Pantheon's divine interconnectedness.

The Delphi Oracle's popularity reached its height in Classical Greece and then continued through the first 600 years of Roman occupation, coming to an end in 392 CE (Current Era), only when a newly Christianized Roman emperor banned all pagan worship and ordered his army to destroy the sanctuary—along with most of the art. The Pythia's last recorded words as the Oracle of Delphi were, "All is ended."

Some 9000 years ago, Gaia's prophetic cave ceremonies marked the opening of a successful 7200-year run. That's impressive. Perhaps there always have and always will be "know thyself" truth seekers willing to travel great distances to seek the counsel of sacred psychoactive truth serums—with or without human support.

It's hard to resist drawing parallels between the frenzied, power-hungry spectacle of the late-stage Oracle at Delphi and the current celebrity-driven psychedelic bandwagon, yet another heart-centered movement being overtaken by industry. And at this pivotal moment, I find the line between "prophecy" and "prediction" especially ironic, and I want so badly to leap into the Psychedelic Pantheon rabbit hole that is Cassandra, the Trojan princess priestess of Troy whom Apollo blessed with the ability to see the future, then cursed so no one would ever believe her—but that's a journey for another day.

To know thyself (as individuals and as a species) is a personal and collective journey and a lifelong endeavor. We cannot cheat the process. Making changes at the level needed requires epic honest inquiry, which requires seeking, finding, and facing truths, even when they are painful.

Truth is the essence of freedom, but all too often, people (including myself) don't want to deal with the tough stuff because it's too hard to bear. We tuck away uncomfortable truths "for later"—helpless in the eye of the storm. Understanding that "the mentionable is manageable" has helped me move through some of the worst of what I'd rather stow away for another day.

To know ourselves as individuals is a worthy quest, but to *truly* know thyself requires knowing self in relationship to others—not just *human* others—and Gaia.

In trip reports, people often wax poetic about how psychedelics helped them "reconnect with nature." I'm troubled by this language. There's still a degree of separation, as if they are in an estranged relationship. These medicines don't *reconnect* us to nature; they help us *remember (know) that we are nature.*

To know thyself truthfully is to know thyself as nature, as animals, as a wildlife species indigenous to earth. Once we *know* ourselves as wild animals, we can begin to decipher which behaviors are hardcoded and which are not—a tricky business—scanning for ideas that hold us captive.

Psychedelic Feminism

For as long as females have been oppressed and sacred psychoactives have been available, there have been psychedelic feminists.

Psychedelic feminism advocates for our natural (universal, fundamental and inalienable) right to journey with sacred psychoactive and psychedelic medicines from the earth in safe, supportive spaces for healing, self-liberation, and empowerment.

We focus on feminist issues because we've been raised female in a male-dominated society so we *know* what that means in real life terms. Violent misogyny is on the rise and our rights are being undermined in nations large and small, and this puts some four billion women and girls at grave risk. That's unacceptable. We don't want these women and girls—or any future generations of women and girls—subjected to what we have endured, what we continue to experience, and what we will most likely be subjected to for the remainder of our lives. We're *NOT* promoting victim consciousness. We just don't want *anyone* to be further victimized by these forces.

Women and girls are so often the target of misogyny that, over time, we each learn to live "on alert." It takes serious vulnerability to find and face the most difficult and illusive truths, to experience and embrace paradigm shifts of perception, to heal from individual, collective and ancestral trauma, and to self-liberate from programming and indoctrination that harms self and/or others. For this to work, we need access to sacred medicines in a *truly* safe, supportive "set and setting" where we can let our guards down.

We know that working on core feminist issues in visionary states can help heal—or change our relationship to—harm we hold as survivors of patriarchy, which includes internalized oppression (a treacherous beast). We are here to heal and self-liberate and to support others to do the same.

Anyone can be a feminist, so anyone can be a psychedelic feminist. Everyone is harmed by patriarchy, which is male-supremacy, which is misogyny. No one is immune. You are welcome here if you're ready to do the work and if you agree to be kind.

Psychedelic feminism is also about helping to bring women's voices to the forefront in the field of psychedelics (and beyond). The psychedelic movement is only as evolutionary as it is equitable, yet males still hold the majority of monetary resources, media machine contacts, and decision-making seats of power, with most females in supportive positions or lead positions funded heavily (if not completely) by male investors or sponsors.

The "divide and conquer" tactic remains one of the most efficient, effective ways to dominate, control and exploit people and has been especially successful in seeding suspicion among people from different "othering" categories, subcategories, sub-subcategories, *ad infinitum*. Cross-pollination and collaboration can be especially threatening to forces that benefit from female subservience. Women from different generations, ancestries, educational and socioeconomic backgrounds who trip together often experience strength in Unity. We learn from one another, and we see and recognize, divide and conquer—in that psychedelic *knowing* way—for what it is. We see, recognize, and perceive the forces that delivered it, and witnessing this together creates rare bonds that can last a lifetime.

We're still served up history with a conspicuously miniscule representation of women because men have owned the narrative for so long. Most of what passes as "evidence" and "fact" is filtered through a lens reminiscent of Narcissus, who fell in love with his own reflection and got lost in it for the remainder of his life. Anyone who has survived a relationship with a narcissist knows how they rewrite history to get their way. This skewed (cringe-worthy) version of history dominates academia, as demonstrated in museums, libraries, and educational intuitions around the world, then spreads to every conceivable dimension of "civilization." (Props to male allies here—we see you.)

With this level of censorship (whether conscious or unconscious), we, *as a species*, cannot "know thyself" for who we truly are, and without knowing who we truly are, we cannot learn from our mistakes. We cannot learn to discern what to hold and what to cherish, what to keep but update, and what to purge. Without epic honest inquiry, we will not get what we need from the sacred medicines and we will not evolve in time to save ourselves from ourselves.

I'd like to see more people (*not just women*) educating themselves about the Age of Patriarchy and the Age of Extinction *as a perfect storm*. I'd like more people to *know* that the force behind the Age of Patriarchy is male supremacy and the force behind the Age of Extinction is human supremacy. I'd like more people to *know*—in that medicine way—that these forces are human-created and human-driven.

I'd like more people to *bring this work into the medicine space* and work on it in the privacy of their own visionary realms, and then I'd like to see them do that again, and again, and again— because that's what it takes. These journeys are not "one-offs." They require brave and steady exploration, maintenance, and community.

I'd like more people (female, male, non-binary, gender fluid—the full rainbow) to bring feminist fundamentals into the medicine space. I'd like to see each of these statements—and many more (herstory is fertile territory)—explored openly and in depth in the medicine space:

Anyone can be a feminist.
Feminism is about the rights of all females.
Patriarchy is not a man or men. Patriarchy is an oppressive construct.
No one alive today is responsible for creating patriarchy.
This is not about matriarchy versus patriarchy.
Women are not a monolith—we are 4 billion unique individuals.
No one is immune to the propaganda, programming and peer pressures of patriarchy.
Patriarchy is a euphemism for male supremacy.

And finally...

I am harmed by male supremacy. Everyone is harmed by male supremacy. All of life on earth and Earth herself is harmed by male supremacy.

Then, I'd like to see these brave and noble psychedelic "know thyself" seekers share their thoughts and experiences, post-journey and through integration—which never really ends—with people they trust and love, who trust and love them, supporting one another continuously and comparing notes.

5

Healing the Roots of Racism

Chaikuni Witan

Chaikuni Witan is a Shipibo Plant Medicine Practitioner, author, speaker and Transformative Men's Coach. Shipibo Plant Medicine is an indigenous healing technology from the Peruvian Amazon, renowned for profound and often life changing transformations.

Since 2012 he has trained in Peru under two highly respected indigenous Shipibo maestros, Enrique Lopez and Ricardo Amaringo. He considers it to be a great honor to be one of a very small number of Westerners trained in this tradition. To his knowledge, he is the only black person with such training. He runs traditional Shipibo style retreats in the US and internationally.

He also runs Men's Retreats, and is in the process of writing a book on masculinity. Both are based the perspective that the masculine and feminine were created to have a balanced and supportive relationship with one another. His work as a healing practitioner contributes to his writing and speaking about race and gender in a manner that focuses on inclusion and compassion, and considers the way trauma has shaped all of our lives.

He has always had unique perspectives on both race and masculinity. The child of a black Jamaican man, and a white American woman, he was raised in the US by his mother in an essentially all-white environment. As he did not have substantive interactions with black people until his adolescence he has always had an outsider/insider perspective on both white and black culture. Similarly, the complete absence of a father, or much in the way of male role modeling, resulted in his coming to his own understanding of masculinity as an adult, again from a very unique perspective.

Prior to working with people, he studied film at NYU and worked for many years as a filmmaker. ChaikuniWitan.com

Perspective and Background

What Does Plant Medicine Have to do with the Roots of Racism?

When the George Floyd protests broke out, I was in the Peruvian Amazon, studying Amazonian Plant Medicine. Amazonian Plant Medicine is a complex and powerful healing technology of which ayahuasca is the best-known element. It is a vast world, with many traditions and lineages. I study with a *maestro* from the Shipibo tribe. The Shipibo approach is known for its depth and precision. It focuses on healing core traumas. In ceremony, each participant receives a healing song specifically designed to address said traumas. It is also my life's calling; I have been practicing for several years and studying for just over a decade.

After the initial shock and horror about what had happened, I was inspired to hear about the surge of humanity fighting against injustice, and even more so, that broader forums were open to the voices of black people and new ideas about how we could all live together. It seemed like a watershed moment in American history. I soon became disheartened at how much anger and polarization rose to the forefront. People had reasons to be angry. Centuries of reasons. But did we really think accusation and recrimination would lead us to where we wanted to go? At the same time, I observed a crosscurrent, particularly amongst white liberals, of confusion, hopelessness, and guilt. I had to ask myself, was this the best we could do?

I began to consider it from a Plant Medicine perspective, which is very different from an activist or structural perspective. There is no good/bad moral framework. There is no accusation, there is only causality and healing. It focuses on compassion, understanding, and connection. It works best one person at a time. And most importantly, it looks at the root of the issue.

Why the Roots of Racism?

People often ask if I can help them with things such as depression or addiction. However damaging these things may be, I do not consider them to be the real issues, but merely symptoms of trauma. I find it more effective to look at the roots of the issues. For difficult traumas and emotions, we literally sing the words "pull out by the roots." Although it can be challenging, if you are able effectively to address these issues, depression, addiction, or any other symptoms abate.

Looking at racism through this lens gives a very different perspective than looking at it through an activist lens. While structural and legal issues, not to mention individual racist actions are huge problems and need to be addressed, a Plant Medicine perspective sees these as symptoms of human issues that plague racist individuals. While it's clear that much needs to be done to ameliorate the impact of racism, I see an effective solution to racism as ultimately coming from addressing the fundamental issue: the racist person themselves.

If you look at racism like a fire, you could say that black people are burn victims. If so, do you want to focus on getting better at treating burn victims or do you want to figure out how to put out the fire? As a black man, I can tell you I want the fire out. Moreover, while legislation is necessary, laws alone will not curb human action. If the anger or fear in a man's heart tells him he must start such a fire no law will stop him. To continue the metaphor, what I am proposing is a way to stop people from wanting to build fires that burn the skin of black people.

Yes, this is an ambitious goal. And yes, I have hope. The hope comes from the transformations I have seen through Plant Medicine. Time and time again, I see people come to retreats with problems that remained unsolvable for many years and across many healing modalities. After doing some harrowing work, I see them leave with healing they had previously imagined to be impossible. It is seeing these transformations, which are often far beyond what is considered possible through Western approaches, that gives me the hope that such a transformation is possible for society at large.

So what does this look like for you? The essence of what I ask you to do is, much like in a ceremony, to look deep within yourself, to accept both your beauty and your ugliest parts, and to see the potential for both destructive and loving behavior in you. From here, you will begin to see how these realities and possibilities exist in every man, woman, and child on the planet. This unflinching looking inward creates the possibility to connect with people whose actions are injurious, monstrous even, and begin to create real change, first in their hearts and then in their actions. This is how we lay the foundation for a more equitable and harmonious world.

While I hope my ideas could be of use in a variety of situations, the primary focus would be in relating to the growing number of dispossessed young white men in America with neutral or slightly racist inclinations whose dimming future leaves them vulnerable to radicalization by white supremacy organizations. That said, although this essay focuses on black and white race issues in the US, these same principles hold true for any situation in which there exists a significant power inequity, be it along ethnic, class, sexuality, or gender lines.

Origins of My Perspective

I am the child of a black Jamaican man and a white American woman. I was raised by my mother in an almost completely white environment. My mother and her friends had moved from a small city to the country to pursue a dream of living in nature, progressive politics, food co-ops, and folk music. Though they were living on modest incomes in rural America, their backgrounds—urban/suburban, middle class, and relatively educated—and their mindset clashed with their environment, which was economically depressed and culturally isolated. It was full of small farms, rusty pickup trucks, and NRA cards.

I was a black child, being raised by a white woman in a white community that was at odds with its surroundings. To top it off, I was an only child in a not particularly social household and had mediocre social skills, which is to say there have always been many layers between me and any community, which means I have always looked into both the white and black and pretty much every community from the outside.

I had the good fortune to experience very little racism growing up. There were a few ethnic slurs here and there but compared to dealing with the embarrassment of the weird food in my lunchbox, not having a TV, and the other seemingly horrific things that separated me from the other kids, they were a nonevent. I am grateful to say that my adult life, which has mostly been in various bohemian communities, has also been fairly untroubled by racism. All this is to say that I also have a very external perspective on racism.

Progressive politics were a big thing when I was a kid. Some of my earliest memories were antiwar songs and protests against nuclear power plants. I brought these values into adolescence. When the first Gulf War broke out, I got on a protest bus to Washington, along with some of my mother's friends. I felt empowered and enjoyed bonding with the adults. We spent the next day marching in Washington. It was an experience: crowds of people, signs, chanting. Against that backdrop, one memory specifically stands out. There was a punk guy: orange hair, painted motorcycle jacket. He was screaming, full of rage. Going to a protest of my own volition was an exciting and new experience, but by the end of the day, it was clear to me that we hadn't done a damn thing to stop the war.

Although I probably didn't have words for it at the time I took two things from this experience. The first is that a lot of what is called political activism is more about a sense of community or empowerment than actually effecting change. The second is that activism attracts people venting anger, which often has nothing to do with the actual issue at hand.

My disappointment led me to the question of how to effect substantive change in the world. I was stumped for some time. A little over a year later, I took my first hit of acid. It definitely changed my world. Psychedelics became a big part of my life and I was sure they were the way to change the world, though I was a little foggy on the implementation. After nearly a two-decade detour through the film industry, I eventually discovered ayahuasca. I was sold after the first cup. Within a couple of years, I began studying Plant Medicine in Peru. Now working with plants to help people is what I do; healing the world one song at a time.

Racism

Racism from a Plant Medicine Perspective

The fundamental thing we need to understand, and that almost everybody misses about racism, is that it is not an unnatural or new phenomenon. Moreover, it is not going away. This is because what we call racism is the shadow of what we call community. It's Sociology 101: in-group good, out-group bad. The healthy expression of this phenomenon is taking care of one's neighbors, pride when the home team wins the game, and your desire to love and protect your family. The unhealthy expression is slavery, war, or choking a man to death because he was accused of passing a counterfeit twenty-dollar bill. The unhealthy end of the phenomenon is aggravated by competition for scarce resources, be they material or emotional.

Competition for scarce resources and favoring familiar communities have existed throughout all of history, in humans, other animals, and plants. Early human remains show evidence of armed conflict. Wolves have packs, which of course defend themselves. When pine needles drop, they acidify the soil, which facilities the growth of pine trees, and inhibits the growth of other plants. Everybody and everything is wired to look out for their own. This will not stop, and it's dangerous, naive, and ineffectual to pretend that it will.

I do not say this to be pessimistic or defeatist. I absolutely see hope and potential for greater harmony between different walks of life. I also know it is predicated on understanding the reality of what we're dealing with so it can be addressed effectively.

Nobody Wants to Be Racist

Racism or any desire to hurt or impugn others comes from a threat to one's need for safety, love, or connection. It is fueled by fear, anger, and other emotions that can be very damaging. Because it defends or gives access to power or resources, people think they like it. The reality is that there is no human on this planet who in their heart of hearts does not yearn to live in harmony with their neighbors. Because of individual traumas, and the false sense of scarcity foisted upon us by a hierarchical economic system, people have been duped into thinking the only way for a good life for them is bad life for other people.

This can be difficult to see for several reasons. If someone is behaving in a way that is injurious to us or others, it can be challenging to see what's actually going on with them. Compassion and understanding are difficult when we feel threatened. Moreover, it would seem that people enjoy doing racist things. Why else would they do them? But who injures another person when they feel safe and loved? If you look at it this way, it's pretty clear that racism comes from pain.

Another big issue is that oftentimes, would-be allies are in such a hurry to support "the cause," and so ashamed of or afraid of anything that could be called racist within themselves, that they rush to condemn and compartmentalize people engaging in anything that could be called racist behavior, so everyone's sure they're on the right team. This is pretty much the opposite of actually seeing and understanding a person's behavior.

What I'm hoping to get you to see here, is that the way forward is to see racist behavior as being like all damaging behavior, in that it comes from a place of trauma, and indicates a person's need for healing and understanding, not condemnation. We need to differentiate between saying a person's actions are unacceptable and taking away their humanity.

We Are All Racist

We are all racist. I say this not to accuse you of not doing enough for "the movement" or failing to refer to someone who has different skin color than you with whatever language is politically correct this week. I am saying everyone has communities they prefer. Everyone has biases. These biases are on a continuum with what we call racism. This is not bad. It is part of being human.

There are many kinds of communities and groups people identify with: tribes, sports teams, red state/blue state, the psychedelic community... When I'm honest with myself, I find that there are some that I feel more comfortable with and feel better towards. Take moment to think about some communities around you, and how you feel about them. I'm pretty sure that if you are honest with yourself you will come to the same conclusion. There is nothing wrong with that. We all find it easier to feel safe, loved, and connected in particular environments.

Early in this essay, I wrote about how when it gets really out of balance: war or slavery. What I want you to consider now is that it gets a little out of balance in us all the time. And that's OK. What I see in a lot of what is called "progressive" thinking is this endless policing of petty transgressions against some antiracist ideal. Yes, we do need to track and attempt to ameliorate our biases. But as a black man, I find the ferocity around this to be damaging: accusatory, guilt-inducing, and most importantly:

divisive. I imagine that for well-intended white people, it must be all those things and very confusing and exhausting as well. Have I scoured the far reaches of my psyche to eliminate any possible bias? It's a zero-tolerance policy applied to the human condition.

We all have biases. I certainly include myself in that. Some of them I've let go, some remain. When I was a kid, I thought the Middle East was full of people with machine guns who wanted to kill America. It's mortifying to write, but I wouldn't say I'm ashamed of it. I see such ridiculous perspectives as a function of the environment I was in as opposed to something about me as a person. Now I live in a much more cosmopolitan world, and think of the Middle East as just another place, albeit with harsher political realities than most. I have a handful of wonderful friends from there and would love to go visit. To me, this illustrates that we all get programmed with certain perspectives when we're kids. Inevitably, some of them were not good. They can change. Having had perspectives that we're no longer proud of doesn't make us bad people. It makes us human.

Speaking of childhood, I did not feel particularly welcomed in the larger social world in which I grew up. If I'm honest with myself, I'll probably always be slightly suspicious of people who wear flannel shirts, drive pickup trucks or drink Miller Lite. This is not likely to change. Nor do I beat myself up about it.

So I ask you to give yourself some space without shame, to any biases, healthy, not so healthy, or somewhere in the middle, that you may have. Maybe write them down—you can burn the piece of paper afterward. Maybe tell a trusted friend. But the main thing is to accept them. Yes, you do want to work to change them, but what I'm saying here is do not condemn yourself for them. This is valuable for you as a person; it's also necessary for a less racist world because in seeing (hopefully) small and not particularly damaging biases in ourselves, we can come to understand that larger and more damaging biases are on a continuum with our own. Seeing the commonality allows us to connect with or at least relate to and see as human people whose biases are actually dangerous.

Divisiveness

It is easy to see racial struggles as us versus them. In an immediate sense, this can be true. If you're in a dark alley with someone who wants to hurt you because of the color of your skin, it's you against them. But in a larger context, this opposition is manufactured by a sense of scarcity. In reality, there's plenty of room for both of you to walk through that alley. If both of you felt safe, there would be no need for either to harm the other. While it's easy to fall into us versus them when threatened, what we need to understand is that ultimately polarization breeds racism. The more we get into the "out-group bad" end of the spectrum, the more enemies we have, and the more people consider us their enemy.

This means, to the extent you want a less racist world, do not look for enemies. If we want to work against racism, we need to be sure we are fighting against racism, not against racist people. The more you push people away the more easily radicalized they are. Polarization was very useful to Donald Trump. Sowing divisiveness and preying on the fears of disposed young white men were fundamental to his or any nationalist political agenda. The CIA describes ISIS using these strategies. It is also contributing to much of the nationalist movement in Europe.

If you feel called to call someone out or condemn them, consider whose goals your actions serve. It is not very useful for those of us who want a fairer and more equitable world. Admittedly, it is easy to get pulled into a polarized perspective. You may feel an immediate and most likely very justified sense of anger or indignation. Making a practice of the work suggested in the previous section can be very helpful for deescalating your feelings towards someone who said or did something you found hurtful. It is hard work. No one is perfect. We all fly off the handle from time to time. But when you consider what's at risk and what the potential is, hopefully, you will find it worth the effort to give a little grace, to accept imperfect allies, to hold out a torch for those who are still lost. Doing so expands your community, and makes the world a safer place.

Connecting

Understand People as Individuals

Dehumanization is the fundamental tool of racism. Simply reducing people to part of a group and allowing that to be your fundamental understanding of them is dehumanizing. Again, there is a continuum: "Jews are greedy," "Mexicans are dirty," "Brits are uptight," "New Yorkers are arrogant," or "Politicians can't be trusted." Presumably, if you're reading this, the earlier statements don't reflect your views, but there's a good chance something later on might. We all dehumanize a little bit. While we're often very vigilant against these thoughts towards disenfranchised groups, we tend to let things slide when considering people who have more power. Generalizations about Jews or Mexicans don't feel good. But politicians? It's open season.

It's also important to understand that sometimes we do this in ways that do not impugn but still lead us down the same path of dehumanizing.

I have a white friend who was interested in working in black civil rights causes. I was very saddened by the reason she decided not to. She felt that black people wouldn't trust her because she was white. Clearly, there has been a lot of mistrust of white people by black people, but to imagine that was the only possible relationship dehumanizes both parties.

I was outside of a dance event and a friend introduced me to one of her friends. The event had a mostly white demographic. The guy was black. He told me that he found a lot of the women attractive but was afraid to talk to them because he figured they thought he was some big dangerous black guy. To me, he was dehumanizing himself as one who was perceived as dangerous—or at least anticipating others doing it to him—while dehumanizing the white women as creatures who were afraid of black men. Obviously, he did not have bad intentions but still created a barrier to people seeing each other as actual human beings, and quite clearly blocked potential connections.

So I ask you to consider times you've dehumanized people—seen someone as part of a group about which you have certain assumptions, as opposed to as a human being. Certainly, try to change, but understand what I am advocating here is not self-policing, but self-awareness. Be aware that you do it. Be aware that we all do it. This a good practice in life and necessary if we want to connect with and change people who are on the malevolent end of dehumanizing.

Own your beliefs. Let other people own theirs. This fosters dialogue.

Many black people feel white people can't have or express opinions about race. Many white people feel shamed into agreeing. This is counterproductive. Fairness requires that white people listen to perspectives and encourage the expression of black people, as our views have been silenced and our needs unmet. Moreover, these views have to be given more weight in public policy to correct past injustices. Fairness and effectiveness also require that white people are encouraged to own and express their own opinions. Space for one's opinion allows one to understand it, which makes it easier to change it when presented with new ideas. Simply being white does not automatically invalidate one's opinions on race. Giving space for their opinions builds cooperation and trust, and lessens resentment.

When dealing with people whose perspectives you find challenging, you will get further swaying their opinion if you let them express and own it first. Although it's tempting to silence or diminish the value of opinions you disagree with, it's more empowering to everyone to allow people the legitimacy of their own perspective. To the extent would-be allies squelch their own perspectives, they will sing along, but their heart will not be in it. This can pass laws, but it ultimately leads to resentment, which will eventually emerge as pathology and animosity. If people who are already trending racist are silenced, they will find the sense of empowerment offered by white supremacy movements more appealing. In either case, listening will ultimately get us a lot further than silencing.

People Need Space to Change Their Perspectives

We all find new things difficult to adjust to. Face masks, the latest iOS update, and new tax regulations. Although these may seem unrelated to race issues, it's the same principle. It just gets a lot stickier when something pushes up against our belief systems or threatens our sense of safety. To the extent that we have new social agendas that we want to succeed, we need to give people who find them threatening a little bit of space to adjust and consider how to present things in a non-confrontational way. It's the only human approach. It's also much more effective. Failing to do this is a big part of why the inspired but unfortunately named "Defund the Police" movement became so problematic politically. It sparked opposition in people who never considered the content because the name was so frightening.

Giving people space for their resistance and time to get used to change takes them from enemy to little brother. "How can we help you get used to this?" goes down a lot easier than "Deal With It!"

Take a little time to think about things you have had a hard time adjusting to. Big, little, political, or otherwise. Think about what would've made it easier for you to take in the new reality. Bear these memories in mind when you find yourself talking to someone who is having a hard time with new ideas or social values that are important to you.

Difficult Emotions in Healing

If these are such great ideas, why isn't everyone doing them? Because we're all totally preoccupied with our own trauma.

Clearly, there are a lot of very intense emotions around race issues and activism. Taking a moment to examine emotions from a psychological perspective is very useful for deepening our understanding of what is happening and seeing how to move forward productively.

People who have been harmed by racism have a lot of very justified fear, anger, and sadness. Fear comes from a threat to self, be it physical or emotional. Anger emboldens, protects, defines boundaries, and gives energy. It is ultimately about not being valued. You clearly wouldn't enslave or beat a person if you value them. Anger and sadness are like twins, they each protect us from the one that is more difficult to deal with. If you find the amount of anger in the world of activism challenging, understand that beneath it there is an even deeper well of grief.

Allies often feel a great deal of shame about the position of black people. Shame is the most difficult emotion to work with. It hides behind the heart, locks it, and prevents healing. It can also be very motivating.

Although change happens by many means, these emotions have a particular synergy to create change in what ultimately turns out to be quite double-sided. Anger and fear motivate. Anger moves people. People hear it. This generates change.

Shame, on the part of allies, can lead to support or acquiescence, which can both be helpful for a political cause. But when you look at the long game or consider more subtle impacts, it gets messier.

If there is some element of guilt or shame fueling your support, although you may feel good about it on one level, ultimately it will never sit well with you, albeit perhaps only at a subconscious level. There will always be some kind of resistance. Eventually, this leads to pushback, resentment, or dissatisfaction. This can push your beliefs in the other direction, again possibly subconsciously. This is extremely dangerous in the long run. Therefore, if you consider yourself to be an ally to a cause that is not your own, make sure you also value your own opinion and needs. If you are fighting for someone else's cause, do so from inspiration, not guilt.

Anger towards others has its own set of challenges. No matter how honestly you came by your anger, it is unlikely to facilitate a person's healing process. This leads to a tricky reality: To succeed, we need to hold the validity of people's — including our own — anger on one hand, and its counter-productiveness to healing trauma on the other. Think about it in your own life. If you're angry at your friend, it won't make them look at themselves. It will make them defend themselves. If it does cause them to look at themselves, it will be because they care about you or value your opinion or love. Given that people engaged in overtly racist activities demonstrably don't care about black people, there is zero traction.

What this means is that if we want people engaging in racist activities to heal, we need to find a way to approach them from a non-angry place. I realize this is a big ask. To do this you would have to work through enough trauma to approach from a non-angry place. I am not saying it's anyone's job to do this. Absolutely not. It is your trauma, and it is your right to relate to it as you please.

I will say that I know from personal experience that such change is possible. I was recently working through some trauma from my childhood. As the process began, and I was becoming more cognizant of what had happened, I wanted to do unspeakable things to the person who had caused it. As the healing progressed, I merely wanted to beat him up. As it progressed further, I felt bad for him and wished he could find his own healing. Although I am unlikely to ever see this man, were I to see him today, I would approach him from a compassionate place.

The more we as a collective have the strength and patience to do this on ourselves, the more powerful of a position we will be in to bring about change in the world. It will allow us to connect with people and hold space for them to change. This is not easy work. It is understandable to be subsumed by one's injuries. At the same time, if we want to make the world a better place it is important to remember the power that is available in doing the work on ourselves so we can stop people from wanting to start the fires that could cause the same injuries to our children.

If this sounds like a pipe dream, consider this: Derek Black is the godson of David Duke and son of the founder of Stormfront. He was the heir apparent to the throne of the white supremacy movement. Several years ago he completely walked away from that movement. Was he called out, harassed, or cajoled? No. He was invited to a series of Shabbat dinners in college. The people there connected with him as a human being and asked him about his life. Eventually, their humanity dissolved the bigotry he had been taught as a child, and he no longer believed in white supremacy. If you are reluctant to examine your anger or hold space for others to change, consider that some college kids with no training at all had the wherewithal to change this man's perspectives, and in so doing, leave the white supremacy movement without its next leader.

Historical Perspectives

It's Ultimately About Class, Not Race

People don't care about keeping black people down, but about making sure someone is at the bottom. The story of Fred Hampton, an often-forgotten figure in the civil rights movement, illustrates this. He was trying to unite poor black people and poor white people. The powers that be immediately saw the danger this posed. He was only active for about two years before the Chicago PD orchestrated the quickest, most ruthless assassination in the civil rights movement. Unlike in other assassinations, law enforcement did not bother to disguise its actions. They hired an informant to map his apartment, drug him, and then in the wee hours of the morning sent a team of 16 tactical officers to take him out. It is a tragic story, but the degree to which his actions threatened authorities illustrates the power of creating allies across race lines, particularly in his case, poor, dispossessed white people—which is to say the people most susceptible to white supremacist mindsets.

The Danger of Shame

The Treaty of Versailles ended WWI on unfavorable terms for the Germans. It also billed them for the war and blamed them for it. The German economy was destroyed. More importantly, the German sense of self-worth was decimated. This cauldron of shame is what Hitler created the Brown Shirts out of. We all know what came next. Consider this next time if you are tempted to shame someone for their actions.

John Lewis

Since the ending of slavery, the most significant advancement for black people in America came from the "Civil Rights Movement" in the early Sixties. Therefore I find it powerful to consider the perspectives of John Lewis, who was Martin

Luther King's right-hand man and sergeant-at-arms. He said that they did not consider it to be a civil rights movement, but a spiritual movement. They considered their oppressors primarily as people lacking in love. This is from a man who was beaten dozens of times resulting in his spending months in the hospital. He trained protestors to look at police, wearing riot gear and carrying batons, as little boys who needed love. Spirit, heart, and conviction allowed them to persevere through endless beatings and arrests, and create an entirely new reality for black people. The success that they had, against an opposition that is far greater than what we face today, is a shining example of how to move forward.

Conclusion

What Can You Do?

You can look into yourself, and acknowledge your flaws, your biases, and your fears. More importantly, find your goodness, your light, and your generosity. Recognize that you and everyone you find wonderful or challenging are human beings, struggling with human issues. Connect with other people from here. Build healthy communities. Listen to people. Seek to understand their perspectives. Validate their reality. Hold space for people whose perspectives you find challenging. It is only from here that you will be able to sway them. Consider inviting challenging people into your communities. Give people time to adjust to new realities.

If you're a white person interested in black causes, learn about black people. Don't just think of us as victims of oppression. It's a drag for you and dehumanizing for us. Learn about our struggles, but learn about our life too. Make friends with us. Ask how our kids are doing, and what our plans are for the holidays. Celebrate our humanity. Take in our culture. Smoke a joint and listen to Miles Davis. Maybe you feel like you should read Alice Walker's *The Color Purple*, as it is a moving story about black people in very challenging situations, but do you know her book *The Temple of My Familiar* is beautiful meditation on love? Ours, just like yours, is a culture to be celebrated, not pitied.

You also have permission to do nothing. There is much to be done, but it is more important that a few people act from their hearts than that many people act from obligation. A spiritual problem can only be addressed through a spiritual solution. If you do choose to act, find faith in the goodness in all people and the potential of humanity, and let this be your guide.

We have all suffered. We have all hurt people. This is true of you, me, and everyone on the planet, even those who have hurt us most. Although it is very easy for these injuries to drive us into animosity, a greater strength and deeper capacity for healing lie in seeing that our pain connects all of us. The way forward is to make space for the healing of all people and all communities. If this is beyond you, hold space for your own healing. It is the most important gift you can give yourself. It is also the most important gift you can give the world because it will create more space for all to heal, most especially those you find difficult. It is in this way that you, as an individual, can open the door to a world in which we can all find harmony, connection, and brotherhood.

6

Psychedelics and Environmental Consciousness

Interview with Dennis McKenna

Dennis McKenna has conducted research in ethnopharmacology for over 40 years. He is a founding board member of the Heffter Research Institute, and was a key investigator on the Hoasca Project, the first biomedical investigation of ayahuasca.

He is the younger brother of Terence McKenna. From 2000 to 2017, he taught courses on Ethnopharmacology and Plants in Human Affairs as an adjunct Assistant Professor in the Center for Spirituality and Healing at the University of Minnesota. He emigrated to Canada in the spring of 2019 together with his wife Sheila, and now resides in Abbotsford.

Since 2019, he has been working with colleagues to manifest a long-term dream: the McKenna Academy of Natural Philosophy, a nonprofit organization founded in the spirit of the ancient Mystery Schools and dedicated to the study of plant medicines, consciousness, intelligence in nature, preservation of indigenous knowledge and a re-visioning of humanity's relationship with Nature. Dr. McKenna is the author or co-author of six books and over 50 scientific papers in peer-reviewed journals.

Interview conducted by Ocean Malandra

∞

OM: Good morning, Dennis. Let's start right at the heart of the matter. Tell me about how you think ayahuasca and other plant medicines are actively changing environmental consciousness.

DennisM: Psychedelics and particularly ayahuasca are the catalysts to open up our reconnection with nature. As a species we are wounded, just look at the mental health spectrum. And really what's happening is we are coming to the head of 2000 years of de-evaluating nature, much of it caused by the Judea–Christian perspective, which says that we own nature, and that nature has no value of its own for its own sake. It exists only for us to exploit and dominate.

I'm a believer in the Gaia hypothesis, which you are probably familiar with. That idea can invoke or sound like something mythological, but if you just stick to the hard science it's clear that the earth as a whole is an ecosystem, the earth as a whole is an organism. And just like all other organisms, it maintains itself through homeostasis. This is maintained through various feedback loops. And these feedback loops we have been disrupting, the consequences of which we are seeing now.

Nature always tends towards homeostasis, but if you push it too far, you reach a tipping point from which it will not be able to recover. And then you get negative feedback loops that push it farther and farther away from recovery. And from where we are now, by the time the people with the power and the money wake up to this, it will be too late. It's almost already too late.

OM: And I know you believe that psychedelics are uniquely posed to help with this "wake up." But I want to ask you what exactly are ayahuasca and the other psychedelic plant medicines? Are they kind of a co-evolutionary thing between humans and the plant world?

DennisM: Yes exactly, that's what I think. That's what I think of all of these psychedelic plants, and beyond that, any plant that we value for food or medicine. They are in symbiosis with us. The plants like to form a symbiotic relationship with us because we will domesticate them, we will take them under our wing in a certain way. And actually, we think we are domesticating them, but really they are domesticating us in many ways. They are directing us to fulfill their needs.

Now indigenous people have this understanding that psychedelic plants, that psychotropic plants, are teachers. They are plant teachers. And we have been learning from these teachers for a long time, but in modern history the tradition has only been kept by indigenous cultures. Now that we are in a globalized world, the plants are spreading from these cultures to the world. In the case of ayahuasca, the plants are escaping from the Amazon to get their message out.

If you believe that the community of species is intelligent and can respond to threats to its existence, one way to look at the rapid growth and spread of ayahuasca is a response to the destruction of the Amazon. It's like we primates, we humans, are the most problematic species on the planet right now. We are able to manipulate technology and forces that have a tremendous impact on the planet. Even to the point of being possibly an existential threat to life on earth, certainly to our continued life on earth.

OM: So do you think that ayahuasca at this point, that there is a kind of insemination of intelligence from the Amazon out into the world?

DennisM: Yes, yes, I do. I think that's exactly what's happening. Plants negotiate their relationships with the environment through chemistry, on all levels. And so they just happen to make these neurotransmitter chemicals that allow them to talk to us. Literally. And what they are telling us is wake up, number one. Several messages, but the big message, the overarching message is symbiosis. Plants want to form symbiosis with our species, that is usually to the advantage of both parties.

There are also several subtexts to what they are trying to transfer to us at this particular historical juncture but the big one is, of course, WAKE UP. Realize the impact you are having as a species on the planet. We as a species need to re-understand our relationship with nature. We have to admit that it doesn't just exist for us to own, to exploit, to rape if you will. Which is pretty much what we are doing right now.

We have been brainwashed into thinking we are not part of nature, but I am here to tell you that we are part of it. And if nature goes down, we go with her. And probably we will go down a lot faster. So, I really think that these plants are trying to get the message out that we really need to change. We have a duty to ourselves and to everything else on this planet to change. We are very clever; we can manipulate technology like no species ever has before to the point that we have all these things at our fingertips, but we are not very wise. Our wisdom does not match our cleverness.

We have to think more clearly about how we use these technologies that have a global impact. We can't just keep doing things because we can without wisdom guiding us. And this is where the plant teachers come in.

OM: How much of this evolutionary flip that needs to happen is related to greed? Because it seems to me that much of the destruction of the Amazon, for example, is not really for survival. I mean just look at the gold mining taking place. You can't eat gold. It's just to get rich. Is greed one of the places where we need ayahuasca to help us evolve our thinking?

DennisM: I think that greed is one of many behaviors that we have that are basically fear based. I believe that greed comes out of fear. The idea that you have to surround yourself with money and all kinds of material things in order to be secure. There is the perception that you need a lot of money to survive. But they never pause to think that this whole system is very fragile, and it can all go away in an instant. And then where would we be? How would money help at that point?

So, we have to re-understand our relationship with nature. We have to be a partner with nature, which is going to require us to become more humble about what we know. There is this idea that we are the pinnacle of evolution and that we have it all figured out, but that's exactly what we need to get over. We are actually the most problematic species that has come along so far. We are also the most promising, but that still remains to be seen. We still need to wise up.

The problem is people tend to not think very clearly about this stuff. The moral dimension comes out of human behavior. There is no such thing as a bad plant. There is no such thing as a bad drug. There are of course plenty of bad ways to use drugs, but that is human behavior. And I think in the conversation that gets lost, that we bring the moral dimension to the table. And that's because we tend to project our moral failings onto external things, it's a way of not acknowledging our responsibility.

Human behavior comes out of the heart. And human behavior can be good or bad, the tools are not good or bad, it's how we use them. That's the important thing to keep in mind.

OM: Do you think it's possible in a way that this global environmental crisis is forcing people to see things from a different perspective? It used to be that you could just drive around all day or throw things in the river all day and not really deal with the repercussions. But now the impact of the global environmental crisis is being felt by everyone with things like climate change and high rates of cancer.

DennisM: Yeah, everyone is impacted by this. But the changes happen rather slowly because in the moment we might be faced with this crazy weather or what have you, but we rarely actually stop and think, "This is climate change, we are seeing it in action." And then you have the deniers. Alternative facts where delusion is the order of the day. These people will be standing in water up to their waist before they ever admit any of this is real. Which is dismaying because these processes occur relatively slowly in comparison to a human life span. But now they are accelerating enough to that we can see them in a single life.

OM: I think you are right. But what I see is at this point the biggest driving force for change is people's own suffering. Things like depression and PTSD are what are driving people in the West to ayahuasca. So, has the plant found a way to meet us at our point of need?

DennisM: It's an interesting thing because the experience that you have of ayahuasca, or another plant teacher, is a one-on-one thing. It's the experience that you have with that particular plant. But the plant also relates to us on a species level, but you can only wake up one person at a time. It's much harder to get a whole culture to wake up. Even though you think that things are spreading, and the word is spreading, it's not clear that it's happening fast enough to turn this thing around.

But that said, what else are you going to do? I have completely lost faith in politics and many other institutions. I have come to believe that plant medicines are really the only thing that may save us. And the question is whether they are getting out fast enough and to the right people. I don't know who the right people are but it would be the people who are in the position to make changes in governments and institutions on a global scale. I don't know if those people are drinking ayahuasca, but I know they need to.

OM: So you are saying we should be trying to change minds that have as much influence as possible?

DennisM: Yes, I do believe that. Let's face it, there are certain people that are in positions that have wide-ranging consequences on a global level. People may say then that you are just catering to the elites. Maybe so, but the elites also happen to control the economic and other forces that can actually make a difference. You got to get the message to these people.

OM: And many of these people are also suffering despite their wealth and power, right, Dennis? Many are experiencing depression and other maladies just like everyone else, no? And ayahuasca and other plant medicines can help them?

DennisM: Absolutely, many elites do suffer just like everyone else. I also know very wealthy people that are elites that have clear minds and open hearts that want to help. Just because you are rich you are not bad. The evil comes in not thinking, or deliberately sort of ignoring what is right in front of our faces. Just willful ignorance, something that in the current political climate is being exalted. We face huge problems, and they are just not going to be solved by denial and keeping people from asking the right questions. We are at a critical juncture, a bottleneck so to speak, and we are not quite sure where it is going to go. So, we need clarity more than anything else.

OM: So how do you think the globalization of ayahuasca, and other plant medicines, is going to play out?

DennisM: Well, what we are seeing now is a huge growth of ayahuasca tourism in places like Peru. And I think it's inevitable although it has both its good and ugly sides. Unfortunately, indigenous people tend to get the short end of the stick when this global culture comes in and says, "Oh you have something cool here, you have ayahuasca, you have a tradition around it. We will take that, thank you, and we won't be giving anything back, thank you."

Now this is not necessarily something to be condemned; it's something that more or less is inevitable. I think that ayahuasca does not belong to just indigenous people, it doesn't belong to anyone. It belongs to itself, it belongs to the earth, and because it is so important as a catalyst for consciousness change, I would like to think that it is the common heritage of mankind.

But on the other hand, it has been under the wing of indigenous cultures until very recently, and therefore they deserve the respect and recognition as guardians of this medicine. But I think what it comes down to is that we are not doing this, ayahuasca is doing this. It is taking over our global consciousness. Because it has to. And it can't happen fast enough as far as I am concerned.

7

Experiencing the Eleusinian Mysteries

Carl A.P. Ruck, PhD, Professor of Classics at Boston University

Dr. Carl A.P. Ruck is a Professor of Classical Studies at Boston University. He received his BA at Yale University, his MA at the University of Michigan, and a PhD at Harvard University.

Dr. Ruck is best known for his work along with other scholars in mythology and religion on the sacred role of entheogens, or psychoactive plants that induce an altered state of consciousness, as used in religious or shamanistic rituals. His focus has been on the use of entheogens in classical Western culture, as well as their historical influence on modern Western religions. He currently teaches a mythology class at Boston University that presents this theory in depth.

He is an authority on the ecstatic rituals of the god Dionysus. With the ethno-mycologist R. Gordon Wasson and Albert Hofmann, he identified the secret psychoactive ingredient in the visionary potion that was drunk by the initiates at the Eleusinian Mystery. In Persephone's Quest: Entheogens and the Origins of Religion, *he proclaimed the centrality of psychoactive sacraments at the very beginnings of religion, employing the neologism "entheogen" to free the topic from the pejorative connotations for words like drug or hallucinogen.*

∞

"Look, I am telling you a Mystery," the apostle Paul wrote to his congregation of Christians in Corinth around the middle of the first century after the birth of Christ, about ten years after he first visited the city and lived among them (1 Corinthians 15:51). As a fully Hellenized Jew, he used the language of Plato to claim that now we see only a puzzling reflection of reality, "through a glass darkly," but "then" [after death] "face to face"; now his knowledge is only piecemeal, but then he will know entirely, with the same knowledge that "he is known to deity" (1 Corinthians 13:12).

In his second epistle to the congregation, as enumerated in the canonical texts, which betray evidence of conflation from several missing originals, he boasts that he has seen with this clarity while still alive (2 Corinthians 12:2 *et seq.*). Fourteen years ago, he had been caught up to the realm of deity, to the paradise garden, whether in the body or not he could not tell, and there he heard unspeakable secrets forbidden to reveal. The paradise can refer only to the Garden of Eden, and the knowledge is that afforded by the forbidden fruit, which made Adam the like of the angels and of God (Genesis 3:22). Half a century later, John, on the island of Patmos, saw a rift open in the ceiling of his cave and he experienced a revelation termed the *Apocalypse* of total truth, normally reserved only for those who had died, the face of God (22:4), and then will be rescinded the prohibition, now to eat freely of the Tree of Life (22:14).

"Unspeakable words" (*árrheta rhémata*) and "mystery" (*mustérion*) were common phrases in antiquity, and especially well-known to the Corinthian congregation, just 50 miles along the coast from the great Mystery sanctuary of the goddesses Demeter and Persephone at the village of Eleusis. The testimony was unanimous that the Mystery was a vision, something seen, but not of ordinary reality. As the votive of Eukrates demonstrates (National Archaeological Museum, Athens, inv. no. 5256), it was something that even a blind man could see, and what he saw was the deity Persephone rising above his sightless eyes. The Roman senator and philosopher Cicero was an initiate, as were most of the prominent politicians, thinkers, and artists, as well as common people of all status in society. By Cicero's time, Athens had become the iconic city of Hellenic culture, and he claimed that among the many excellent and indeed divine institutions that the city had offered to the world, none ranked higher in his opinion than the Mystery (*Laws*, 2, 14:36).

At Eleusis, the vision was accessed by the drinking of a special potion, a "mixed" drink called the *kykeon*, and the evidence about the other Mysteries of antiquity always implicates an intoxicating sacrament for inducing altered perception and a state of rapture and ecstasy. At the sanctuary of the Kabeiroi outside of Thebes, it was the custom to smash the bowl from which the drink was drunk to prevent its ever being employed again for profane use. At the sanctuary of the Great Gods on the island of Samothrace, the vessels were marked property of the gods, and the huge number unearthed in the excavation has led the archaeologists to conclude that drinking to the point of excess was an element in the ceremony. The scenario for these ancient Mysteries programmed the initiate to die while still alive and to survive reborn with a personal communion with deity, the knowledge otherwise afforded only to the dead. The Mysteries imparted the infinite knowledge known to god.

The face-to-face encounter was exactly the way that Plato, who was without doubt an initiate, defined the Mystery. "The whole soul throbs and palpitates" with fear and awe, and one becomes totally intoxicated with the excruciating erotic drive for orgasmic union with the deity (Plato, *Phaedrus*, 250e *et seq*.). Plato's account is wryly ironic, with the initiate experiencing the painful ornithological metamorphosis into Eros himself, because although Plato uses the language and scenario of the Mystery, he had become suspicious of the dangers of intoxication, of visions caused by something one has eaten (Plato, *Laws*, 637d *et seq.*). He substituted a rigorous and lengthy meditation upon mathematical paradigms to induce the final vision of the ultimate knowledge, which is simply that this ordinary reality is only a deceptive imitation of something else (Plato, *Seventh Epistle*).

Such knowledge is called *Gnosis*. For Plato, it was "recognition" since it was already there latent in the soul, but forgotten, needing only to be awakened again into clarity (Plato, *Meno*). Gnosticism as a religious tradition in many faiths refers to mystical or esoteric knowledge derived from direct personal participation with the divine. Many sects in the earliest tradition of their religion induced such communion by a sacred psychoactive food, later restricted for the ecclesiastical elite, and officially deemed heretical and proscribed for the commonality, for whom a symbolic nontoxic surrogate was substituted. In Judaic tradition, *Exodus* (30:22–33) records the recipe for preparing the holy anointing oil of priestly ordination and the incense for burning in the Tabernacle, which contains large quantities of a substance correctly identified now for almost a century as cannabis, ending with the admonition that any unauthorized use should entail that the perpetrator be disavowed by his father's kin.

Paul, in the same *First Epistle to the Corinthians*, reprimands the congregation for abusing the sacramental Eucharist, so that quite a few of them have become sick and some have even died (Paul, 1 Corinthians 11:30). A symbolic meal of wine and bread cannot cause death. He stipulated that anyone who consumed the Eucharist without discerning that it is consubstantial with deity destroys the sacrament and himself. The term now used for such a substance consubstantial with deity is an entheogen, which if abused is just a drug. Archaeological evidence suggests that for certain sects of early Christianity as late as the fourth century, the entheogen was not a symbolic or magically transubstantiated, innocuous surrogate (Basilica of Aquileia, Italy), and several masterpieces of Renaissance art (Grünewald's *Isenheim Altarpiece*, Titian's *Bacchanal of the Andrians*, van Eyck's *Ghent Altarpiece*, Botticelli's *Venus and Mars*, etc.), as well as the decoration of numerous churches (Basilica of San Vicente, Ávila, Spain, etc.), indicate that the tradition continued among some elite patrons and ecclesiastics as the popular science of alchemy, seeking the elixir of knowledge.

Cave drawings as early as the Paleolithic (Chauvet Cave, France) indicate that entheogens were the primordial stimulus for humankind's awareness of the spiritual dimension of reality, what might be termed the First Supper, of which the Last is a commemoration. Such is the antiquity of the motif of transcendence as liberation from the cave of delusionary appearances, which Plato enunciated in his Allegory of the Cave (*Republic*, 508b–509c), applying it to the no less delusionary appearances of ordinary reality. The Mystery sanctuaries were caves or buildings constructed in the symbolism and likeness of caves, and cave incubation was the source of divine knowledge. The Cave of Euripides on the island of Salamis has been excavated. It was here that the tragedian playwright communed with the Muses, supposedly writing his plays. It is a dark, dank, and dreary place, in use as a sanctuary since the Neolithic, and in no way suitable for writing, but appropriate for visionary inspiration. It became a hero shrine and destination for tourists in the Roman period.

Pythagoras, who had been initiated in Egypt into the psychoactive *haoma* sacrament of the Achaemenid Persian Magi, saw a vision of the geometric theorems from a sacred cave, either on his island of Samos or in Southern Italy, where he fled, seeking political asylum and convened groups of his disciples. A cave is an implausible place for a schoolroom. As the poet Ovid described it (*Metamorphoses*, 15.62–64): "Pythagoras with his mind journeyed to the gods far away in space and what Nature denies to human vision, that did he drink down with the eyes of his intellect." The metaphor of drinking the vision is not without meaning. Paul used the same metaphor in declaring the Mystery of Christianity: "A drink has conquered death." (1 Corinthians 54: *katepóthe ho thánatos eis víkos*).

The Persian religion spread as a Mystery throughout the Roman Empire as Zoroastrian Mithraism and it orchestrated a sevenfold scenario of initiation in small subterranean chambers as symbolic caves, beginning with an enactment of the sin of incarnation and culminating as cosmic transcendence accessed by the ingestion of a sacramental meal of bread and wine, symbolized as bits of flesh from the slaughtered Cosmic Bull of the constellation Taurus. The vessel for drinking the sacrament figures in the decoration of the Mithraea. The initiate burst from the cave, transported in the spirit to the rim of the cosmos, from which vantage he saw the totality of existence (Cicero, *De re publica*, *Dream of Scipio*, his substitution for Plato's Myth of Er).

Not only were entheogens at the origin of primordial religion, but newly founded sects in the recent history of the eighteenth, nineteenth and twentieth centuries indicate the same impetus, later lost, rejected, or strenuously denied by those who followed in the wake of the founders. The Ephrata Cloister in Pennsylvania was established in 1732 by German alchemical Rosicrucians. They practiced a secret initiation that lasted 40 days of sequestration, physical mortification, and dietary restriction, at the end of which, aided by Johann Beissel's elixirs, they were conversing with angels.

Helena Blavatsky, the founder of the Theosophist Society, considered opium and other drugs an aid for revelation. The official hagiography of Joseph Smith, the founder of the Church of Latter Day Saints, recounts his discovery of the golden plates inscribed in "reformed" Egyptian hieroglyphs with the *Book of Mormon*, while digging for hidden gold. It is less well known that such digging was a metaphor for seeking medicinal roots. He was frightened at first to touch the plates because a giant toad was using them for its stool. While he was still alive, visions were the common experience in the Church services and in his School of Prophets, which first met in 1833, but the secret of his elixir, and the materialization of angels, passed with him into the grave.

The founder of Scientology was a member of a group that experimented with psychoactive agents to access communion with Egyptian deities. Both Churches condemn the use of drugs. The use of psychoactive agents for spiritual enlightenment was not declared illegal until the Controlled Substances Act of 1970. The Religious Freedom Restoration Act of 1993 reaffirmed the right of indigenous cultures to their traditional entheogens.

R. Gordon Wasson's *Life* magazine article of 13 May 1957 recounting his experience with the indigenous Mazatec shaman María Sabina in the remote village of Huatla de Jiménez in the central Mexican highlands of Oaxaca first brought such traditions to public awareness and precipitated what came to be called the Psychedelic Revolution. It led to widespread experimentation and abuse of drugs like LSD, which the Swiss chemist Albert Hofmann had discovered in 1943, but it was not generally known until his collaboration with Wasson. Within ten years, *Life* magazine reported in 1966 on LSD as "the exploding threat of the mind drug that got out of control."

In his autobiography, Hofmann called LSD his problem child. Its potential for spiritual enlightenment influenced a new awareness of humankind's relationship to alternate dimensions of reality, separated from normal waking consciousness, as William James had written in his 1902 *Varieties of Religious Experience*, by "the filmiest of screens... potential of consciousness entirely different." This new awareness was responsible for the innovative agility that developed computer mentality and discovered fundamental advances in biological science, such as

DNA and PCR (the polymerase chain reaction). James reached this conclusion by self-experimenting with psychoactive agents in the manner common among the elite spiritualists of the nineteenth century.

Huston Smith, who died in 2016, was a participant in the Good Friday Experiment with psilocybin at Boston University in 1962. He became a noted scholar in comparative religion and adept in the techniques of Eastern meditation, but he concluded that he had personally experienced the mystical vision that he had previously known only from descriptions in numerous religious traditions. The same experiment has been repeated more recently at Johns Hopkins University, with the same results reported. Smith wrote of the experience in *Cleansing the Doors of Perception: The Religious Significance of Entheogenic Plants and Chemicals* in 2000, borrowing the title of Aldous Huxley's *Doors of Perception*, published in 1954, in which he recounts his experience with mescaline. The "doors of perception" comes from William Blake's 1793 poem, *The Marriage of Heaven and Hell*, in which he uses the metaphor of Plato's Cave:

If the doors of perception were cleansed, everything would appear to man as it is, infinite. For man has closed himself up, till he sees all things thru chinks of his cavern.

Hofmann came to believe that visionary experience was a valid mode of discovery, analogous to the rigors of scientific empirical evidence. Shortly before his death at the age of 102, he wrote:

Only a new Eleusis could help mankind to survive the threatening catastrophe in Nature and human society and bring a new period of happiness.

Looking back at the end of the century whose mentality he probably more than anyone influenced, he saw the crisis that we humans have created by our destruction of our planet Gaia and the possible extinction of our species.

He was not proposing that ancient Mysteries be revived as religions, but that mankind again be afforded the experience of confronting infinity. Infinity by its nature is inexplicable from a finite perception, but it shatters the confines of ordinary reality. It can assume many faces for the face-to-face encounter, none of them complete. Classical mythology has provided numerous psychiatric paradigms for understanding the psyche or soul of mankind. The ancient wisdom of the story, the myths, that programmed the initiates offers a potent pathway for personal discovery and personal enlightenment. Ultimately, at the end of the never-ending journey, the face that awaits the traveler is one's own.

Topography determined the sanctity of the site for the Mystery experienced at the sanctuary of Eleusis. The alignment of the acropolis with the twin-horned peaks on the mountain ridge to its west identified it as the vulva of Gaia herself. The acropolis was surrounded by the fertile Rarian plain, where the goddess first taught the art of agriculture. The natural secure harbor afforded the island of Salamis lying close offshore invited commerce with the fruits of Gaia's bounty. These gifts of Nature have led to the exploitive abuse of one of the most blessed places of the planet. The sanctuary was desecrated by the Christians, who replaced it with a church of the "Holy Lady Who Resides in the Seed of Grain" constructed upon the acropolis.

The Industrial Revolution exploited the harbor, building oil refineries along the shore, an iron foundry, and a cement factory at the edge of the temple complex. Petroleum from oil seeps in antiquity was considered the menses of Earth, and ores and stone were her bones. The refineries and ships have polluted the air and the waters of the bay, and the cement factory has dug into the rock of the acropolis, and most of the marble remains of the temple complex and its art have been converted into cement.

The New Eleusis that Hofmann envisaged has become the Gaia Project promoted by the Friends of Elefsina, the modern name of the village. It will include a new Museum Complex and Art Center for renegotiating the Covenant with the spiritual entities of the planet, who were once seen in the faces of the goddess Demeter and her daughter Persephone. It will be the world center for investigating the wisdom of the old myths. In mythical tradition, the affront to Gaia's sanctity in sequestering the cultivated field for agriculture from the natural wilderness required retribution.

The act of plowing was a common metaphor for the sexual intrusion of the plowshare into the vulva of Earth and for the begetting of children. The ancient tale told that the first plowman, who was the lover of the goddess, was offered as a victim. Each year, the rite was commemorated by the inaugural plowing of a sacred plot, in honor of the first plowman, and an animal was slaughtered in his name.

The entheogen allows one to pass the interdimensional barrier and see that what one thought was a myth is a reality. One must give back recompense for what is taken. The Center will also investigate the new technologies for environmentally neutral modes of survival upon the planet Earth. The future, if it is to exist, will start at Elefsina.

Section 2

Societal Transformation

8

Plant Power to the People

Larry Norris, PhD, Co-Founder of Decriminalize Nature

Larry Norris, PhD, studied biopsychology and cognitive science as an undergraduate at the University of Michigan, and defended his doctoral dissertation at the California Institute of Integral Studies (CIIS). His dissertation research reviewed archived ayahuasca experience reports to identify transformational archetypes and insights that could help inform developing models of integration (meaning-making).

Larry is the co-founder and executive director of Entheogenic Research, Integration, and Education (ERIE), located in Oakland, CA. ERIE is dedicated to the development of community education, research, and integration models related to entheogens, and is currently offering workshops to cities across the US who are working to decriminalize entheogenic plants and fungi. Larry is also a co-founder and board member of Decriminalize Nature (DN), which sprouted from Oakland in 2019. He advocates for the unalienable right to develop one's own relationship with Nature and aims to support efforts to decriminalize entheogenic plants and fungi (e.g., ayahuasca, iboga, cacti, mushrooms).

∞

Can you feel it? There is a battle going on for the agency over consciousness. We are presently witnessing an alarming degree of medicalization and corporatization of psychedelics. Gatekeepers and profiteers are developing regulations to limit access to sacred consciousness-expanding plants and mushrooms. Monopolizing patents, cost-prohibitive therapies, costly licensing, conflicts of interest, and surveillance capitalism are all too common. Despite this, a powerful grassroots campaign called Decriminalize Nature (DN) is gaining traction across the United States. DN emerged to add another voice to the discussion and devise a strategy that would always include the people. In DN's view, decriminalizing all natural psychedelics (entheogens) must happen without sacrificing community rights and individual autonomy in favor of corporate profits and governmental controls over personal sovereignty.

DN represents the principles of indigenous, grassroots, and underground communities existing before pharmaceutical firms and profiteers decided to participate. DN recognizes that sacred plants and fungi teach us new ways to live in the world. Ways of being that go beyond mental health and help us comprehend what it means to be a human being on planet Earth. DN is a modern manifestation of people honoring ancestral knowledge and practices that seek to address the many issues we face in an equitable and holistic manner. Consciousness is far more extensive than the material world we live in, which may terrify a system that relies on fear and disempowerment to feed consumerism and imperialism. These are not simply medicines for the mind; they are also medicines for the spirit. These spiritual medicines are the keys that open doorways to accessing higher states of awareness and interconnectedness. It is time to liberate consciousness by changing policy and taking back our sacred right to develop a relationship with nature.

But not all approaches to policy lead toward individual and community liberation.

Medicalization, legalization, religion, and decriminalization/ descheduling are some current ways to combat the Controlled Substances Act and increase access to entheogens and psychedelics. Each have their own value, but the larger psychedelic community is concerned about who is crafting the policy and how they will implement it. The corporate control of the medical and legal approaches is certainly unnerving. However, there is another option. One that maintains our historical connection to nature while allowing for a community and economic exchange that stays local and grassroots through a grow-gather-gift model. So, how should we proceed? We must learn about the history, flaws, and promises of current policy, and then advocate for the people and the sacred.

Decriminalize Nature's Vision and Values

Decriminalize Nature (DN) was founded in December 2018 to fill policy gaps and restore our sacred relationship with nature. Decriminalize Nature, at its foundation, "recognizes that the current grand struggle of humanity is a battle of paradigms — chief among these battles is the paradigm of scarcity, fear, and competition vs. abundance, compassion, and cooperation" (DN, 2022). We started by asking each other what was missing from the present policy and what we could bring to the table. We looked at the possible difficulties and benefits of impending legislation in Denver (2019) and Oregon (2020) and whether we could modify policy in Oakland. While discussing our ideals and approach, it became evident that there was a niche that could inspire the people by highlighting, embodying, and fighting for concepts like nature, the sacred, the grow-gather-gift model, and empowering individuals and communities at the grassroots.

The DN board of directors issued a statement explaining its position on indigeneity, sacred plant medicines, and sustainability. This fourteen-point declaration recognizes the advantages and inventions of the Western worldview, but it also acknowledges the history of Western exploitation and extraction of resources that hurt the earth and each other. According to DN, this time in history requires radical indigeneity, "where answers to the world's largest social and environmental crises must incorporate compassion, inclusion, and reverence for all of nature and nature's creations, including our fellow humans..." (DN, 2022). With 60% of the DN board having indigenous heritage, DN values the Indigenous worldview. This worldview statement advocates for approaches to entheogens that seek to educate, advise, and assist rather than condemn, prosecute, and imprison. DN urges everyone to trace their ancestors and adopt an indigenous worldview "regardless of skin color, ethnicity, race, cultural background, or religion." To assist the emergence of consciousness, we must all reconnect to our roots and guarantee that the corporate and pharmaceutical industry does not conquer the final bastion of consciousness.

The relationship between humanity and nature is one value DN mentions emerging from the Indigenous worldview. We are nature and criminalizing our interactions with it criminalizes us. In order to collaborate with all entheogenic groups, DN chose to focus on natural plants and fungi on schedule 1: Psilocybin and psilocin-containing mushrooms, mescaline-containing cacti, DMT-containing plants, and plant combinations such as ayahuasca, and ibogaine-containing plants. Given the existing pharmacological exceptionalism with poppy and coca, which are both listed in Schedule 2 on the CSA, DN did not address these natural substances. We know that entheogens are sacred earth medicines utilized by our ancestors' cultures across the globe for centuries, if not millennia. Many DN members have also shared stories of how entheogenic experiences helped them reconnect with the environment and the planet. Respecting and honoring entheogenic plants and mushrooms that arise from the earth is a necessary step in healing in an age of environmental disaster. Thus, DN seeks to honor these entheogenic plants and fungi, as well as the earth-based traditional practices that use them for ceremonial or spiritual purposes, and believes they should be cherished and preserved, not criminalized.

While DN agrees that no one should go to prison for using any substance, we did not include synthetic materials. We realized that the most powerful thing we could support was natural plants and fungi that individuals could cultivate, eliminating the need for them to rely on a pharmaceutical company or laboratory to access their experiences. We are already witnessing patent-jockeying with synthetic materials, but nature cannot be patented, and no one owns nature. While some parties are attempting to claim ownership of specific plants, we do not believe anybody should go to jail over them. Claiming ownership over plants or patenting synthetics are comparable in that one group seeks to monopolize and control access to the Creator's gifts. A paradigm of colonization would advocate for incarceration as a punishment rather than educate and share techniques that might aid in stimulating the development and appreciation for these sacred plants and mushrooms. If humanity is to have a fighting chance of exploring its awareness and promoting personal and spiritual growth, we must elevate the most accessible options.

We must revitalize a worldview that acknowledges that nature is sacred and there is a long history of using entheogens in communal ceremonies and culturally relevant settings. When writing the resolution, we discussed many ideas for ensuring the sacred and spiritual framework stay embedded in the policy. We chose the term "entheogen" (generating the divine within) over "psychedelic" (mind-manifesting) to better reference natural materials and ancient practices.

As evidence of the mainstreaming of this conversation, the term "psychedelic science" now refers to clinical approaches, chemical compounds, mechanistic neuroscience, and mental health pathologies. But, as we know, nature-based spiritual practices have long existed outside the realm of conventional science and psychology. While the term "entheogen" can apply to a range of psychedelics, both synthetic and natural, it has recently come to be known to refer to natural plants and fungi. For the sake of this article, psychedelics shall refer to corporate, clinical, or pharmaceutical approaches, and entheogens will refer to sacred, natural, and community-based approaches. These competing opinions highlight the conversation's sacred/ secular split. According to DN, "sacred plant medicines provided by Creator, and stewarded by all of our ancient ancestors who walked this planet for thousands of years before colonization, industrialization, and commodification, are portals enabling any people who approach these medicines with reverence to connect with Spirit and the Divine and to find their way back home to the indigenous worldview" (DN, 2022). We respect the Indigenous worldview, viewing these plants and mushrooms as sacred and seeking harmony with the planet. Whatever one thinks of entheogens' spiritual potential, what could be more sacred than liberation and sovereignty over consciousness?

DN aspired to go beyond a policy that addressed use and possession and incorporate cultivation and community-based sharing and practice. The "grow-gather-gift" concept is an idea that encourages people to grow their own, gather in a community (or sustainably harvest in nature), and share amongst themselves. This concept would not only give people access to entheogens but also foster a stronger connection with nature, promote sustainability, and encourage the development of community-based projects. We are encouraging and teaching people that they can grow their own entheogens or, if they cannot, share in the community with others who have cultivated more than they need. This paradigm would provide access to everyone who wants to engage with entheogens without relying on pharmaceutical firms, pricey luxury retreats, or over-marketed and packaged dispensaries. Whatever happens with medicalization and legalization, we will always have access to these sacred natural materials if we can cultivate our own plants and mushrooms without fear of imprisonment.

Our mission is to enable ways for communities to cultivate entheogens, congregate in their communities, and teach people how to interact with elected leaders and effect change. To do so, we needed to establish a narrative in which everyone could participate, and the movement's expansion demonstrates the desire of the grassroots community to fight for their rights. Collaborating with numerous communities around the United States has been a privilege. Despite allegations of a lack of diversity in clinical models, powerful women and people of color lead most DN teams. While our efforts are focused on entheogenic policy, hundreds of individuals around the country now understand how to interact with their elected officials and enact policy that can effect change at the local level. We are committed to winning one city at a time while teaching our elected leaders about entheogens by discussing compassion, abundance, human rights, scientific evidence, and ancestral usage. We must keep in mind that elected representatives serve their citizens. So, we must ensure our message is heard by connecting with them directly and expressing how entheogenic policy will help their constituents. Furthermore, while we should not need permission to connect with nature, cultivating favorable ties with political leaders can only benefit the cause.

Decriminalize Nature is more than simply a policy organization. We use policy to inform people across the country about the potential benefits of entheogens. We evolved as a group that emphasized the value of these entheogens not simply for mental wellness but also personal and spiritual growth and the exploration of consciousness. This policy is a human rights issue; no one should be imprisoned for using, possessing, cultivating, or partaking in entheogenic practices. The phenomenal expansion of the DN movement across the country implies that others agree. On the other hand, DN lives amid a sea of different approaches to policy, each with its own objective, some more equitable than others.

Policy Approaches

The terms "legalization" and "decriminalization" are sometimes used interchangeably. Many individuals have asked us about the difference between these approaches. The following section briefly explains the distinctions between the three primary models, in order of strictest regulations: medicalization, legalization, and decriminalization. The religious exemption is another factor to consider, although it is outside the scope of this article. It is worth remembering that all aim to respond to criminalization in various ways. However, each model serves various groups differently, so understanding the context is critical when determining whether to advocate for these approaches.

The FDA must approve the medicalization approach, where clinical practitioners will deliver psychedelics. Medicalization is limited to mental health issues such as treatment-resistant depression, end-of-life anxiety, post-traumatic stress disorder, and substance abuse issues. If authorized for medical use, there would be a rescheduling of these substances. While this may slightly lessen the criminal penalties for non-medical people, it does not make them legal for anybody to use. Only those in the medical field are permitted to administer them, and only for mental health conditions with tight inclusion and exclusion criteria. The medical industry, in this example, corresponds to pharmaceutical companies that will profit from supplying the synthetic product to clinical settings. While there are instances where a clinical model can be beneficial in treating various ailments, this should not come at the price of criminalizing people who do not meet the study's inclusion or exclusion criteria or cannot afford the therapy.

Legalization is a state-controlled, highly regulated approach requiring expensive licensing and training. Some have referred to legalization as "prohibition lite," implying that it is not truly lawful unless adhering to the state's costly and stringent rules. Using the term "illicit" and prosecuting those for cultivating a plant or mushroom, whether inside or outside the legal framework, is ludicrous. Often these state licenses and permits are expensive and only available to those who have large amounts of capital. In Oregon and Colorado, corporate interests are developing expensive luxury retreat centers for therapeutic purposes. Even outside the luxury retreats, some estimate that an individual in Oregon will pay between $3000 and $5000 for treatment. Because of the high expense of these treatments, there is already discussion regarding insurance to cover the costs. We anticipate that if insurance regulators control access, they will prefer synthetic psilocybin over natural mushrooms because of the desire to reduce uncertainty and deviation of the experience. Time will tell if this is an accurate assessment. Currently, legalization does not include dispensaries, and while a dispensary had opened in Oregon while they were waiting for the implementation of M109, they were quickly raided and shut down. As research into the therapeutic effects of psychedelics continues, we can only speculate on what role they will play in the future of mental health care. However, these ancient entheogenic healing modalities have the potential to bring profound changes to the lives of millions of people.

Decriminalize Nature's goal with decriminalization is to remove criminal penalties or punishment for entheogens. If mushrooms, for example, were never against the law, the government would not need to legalize them, just like it does not need to legalize air, water, and sunlight. We must remove the criminal penalties and take back entheogens for the people. Some believe decriminalization must still require a penalty, such as a monetary fine or mandatory drug court. However, this does not need to be the case, as we have seen in cities across the US. On the municipal level, this looks like deprioritization of law enforcement and restricting any funding toward investigation, persecution, or arrest for entheogens. At the state and federal level, there can also be a descheduling or exemption from the Controlled Substances Act, which removes entheogens from being a crime. The ultimate goal of decriminalization is to allow access to those who may benefit from their use and help break the stigma surrounding them.

Legalization or medicalization without decriminalization first leads to incarceration and continued penalties. We are seeing this occur in the cannabis industry, where corporate cannabis license holders are now seeking to eradicate home growing in various states to reduce competition and create greater scarcity for consumers. Without decriminalization and driven by business competition, the regulated frameworks can utilize the arm of the law to make the public buy their product and penalize those without state-approved licenses. While some individuals will find any of these approaches valuable to them given the circumstances, we advocate for the decriminalization and descheduling of entheogens with a grow-gather-gift model where local economies can emerge for services in the community, such as guiding or cultivating. This model allows for an inclusive and mutually supportive form of economy, which helps to reduce systemic harms caused by the war on drugs, reduces poverty, and encourages personal and spiritual growth. Of these approaches, we saw decriminalization as the best way to ensure equitable access and avoid the systematic problems emerging from these other models.

Systematic Issues

These plants, fungi, and compounds are classified as Schedule 1 substances under the Controlled Substances Act, which means they are highly addictive and have no medicinal use. The Nixon administration criminalized these substances for political reasons rather than scientific ones. Numerous studies have demonstrated that these drugs are not addictive and can aid in the treatment of substance use problems. However, owing to the CSA's structure, the onus is on psychedelic researchers to demonstrate medicinal usefulness in order to reschedule. Rescheduling, on the other hand, puts these ancient sacred medicines in the hands of pharmaceutical firms rather than the people. The CSA is a criminalization policy that gives corporate interests the keys. The CSA process provides a direct line to big pharma to prove medical value and they must set up Good Manufacturing Practices for synthetics and trials. And in order to do that, pharmaceutical companies must invest large sums of money and resources in clinical trials, thus making them prohibitively expensive for many people.

The CSA inherently creates a false dichotomy between medical and recreational, which became apparent in the cannabis industry. With only two options, medicalization and recreation, this dichotomy privileges the medical system as legitimate while everything else is considered high risk, fooling around, or partying. However, this does not consider all the other reasons why someone would desire to have these experiences that are not deemed recreational by current standards. We need to go beyond the medicinal and recreational applications of plant and fungal medicine. What about sacred experiences, personal and spiritual growth, consciousness exploration, ancestral practices, mystical epiphanies, or creativity, to name a few? While non-medical people are stigmatized in this dichotomy there is nothing recreational about personal and spiritual growth. On the contrary, it takes hard work. Therefore, it is important to recognize the diversity of applications and benefits associated with plant and fungal medicine.

Due to the enormous cost and commitment required to conduct FDA studies, there is now a race for intellectual property to improve the company's worth and provide a return to investors. There are many patent applications relating to psychedelic therapy and processes. COMPASS Pathways, for example, has patented one technique for producing psilocybin. Journey Colab is seeking to obtain a patent for mescaline for the treatment of substance use disorders. If their trials are successful, even MAPS Public Benefit Corporation will have exclusive rights to utilize MDMA for PTSD. How could these corporations patent and control these natural substances in synthetic form and sell them back to us while advocating for natural drug criminalization? Since this material is extracted from nature it should be part of the public commons. This strategy is not limited to entheogenic plants and fungi. Herbalists and others are subjected to the same FDA pressure, which ignores the promise of affordable natural therapeutic substances in favor of synthetics. Patenting and ownership claims are one example of the systematic problems, which might be disastrous for psychedelics and healing.

A psychedelic renaissance implies a resurgence of something we lost (communal psychedelic use). What we are getting instead: Legal sessions that cost three months' rent, patents for holding hands during therapy, and psychedelic compounds with no psychedelic action.
– The Archaic Revival

Another issue with psychedelics in medical contexts is that they exist within the dominant mechanistic worldview. Following an experience with psychedelics, a person may become susceptible or hyper-suggestible. Especially with psychedelics it is important to understand the contextual paradigms these treatments exist in and who is the one making meaning of the experience. Consider having a spiritual experience at a medical clinic. Can the medical industry's mechanical perspective accommodate these spiritual encounters positively? There is a fear that a medical model approach may ignore the spiritual or immeasurable aspects of the experience, which may be crucial to the healing process. We must use caution and prepare a suitable container for the transformative and spiritual nature of these experiences.

There are some cases where the medicalization approach will be beneficial, but the present narrative around psychedelics places too much focus on mental health. Although research suggests they can benefit people suffering from treatment-resistant depression, they are not for everyone. And we must recognize that cultural or societal settings frequently influence mental health issues. If someone is depressed and returns to a depressing environment, they will not adequately address their mental health. Furthermore, these methods for improving mental health do not tackle the underlying causes of depression. Consider the consequences of opening to the enormous expanse of awareness, witnessing what may be, and then returning to a world of struggle, poverty, authoritarianism, income disparity, racial strife, and other societal ills. Wouldn't that be depressing? Indeed, it would be disheartening for many, and such methods may have limited effectiveness in addressing the root cause of their depression.

Medicalization and legalization have high costs for these experiences, which reduce accessibility for low-income and marginalized people. On the other hand, decriminalization has minimal costs. Within a decriminalization paradigm, the commerce question focuses on the services rather than the substances. While some may see this as a gray market, it permits competent individuals and groups who have been stewarding these medicines for decades to give services to their community while still being able to pay their bills. Recognizing the importance of the economy concerning entheogenic healing, DN is establishing a go-local strategy that preserves value creation—both therapeutic and economic—in the community. We believe this will help sustain a vibrant and responsible ecosystem of service providers while furthering economic growth in marginalized communities.

Policy Issues and Considerations

Over the last four years, considerable policy has been written, proposed, and voted on, but is all policy good? The New Approach PAC, funded by deep pockets and corporate interests, is currently the primary driver of statewide policy. However, their track record with cannabis legislation shows legacy markets left out in favor of corporate cannabis. This history is repeating in Oregon and raises questions about the motivations of this PAC and whether its primary focus is really to create sound policy for the people. The policy is often purposefully complex, making it difficult for laypeople to grasp. But we need to stop being apathetic and pay attention. We are talking about the liberation of consciousness. What are the things to watch out for when policy moves ahead in other locations when choosing whether to advocate for it? Limitations, closed task forces, data collection and privacy, exorbitant expenses, and governmental control are just a few examples of concerns.

One area where we must fight for policy change is the removal of limits on natural plants and mushrooms. However, policymakers frequently use legal jargon to conceal their objectives regarding restrictions. Terms like "allowable amounts," "amounts sufficient to," and "personal amounts" all limit an individual's access to entheogens and local economies. Even the term "allowable amounts" affects the power dynamic, implying a need to seek permission from the government rather than taking back these sacred medicines for humanity. There is never any mention of limitations for corporate or pharmaceutical interests. Nature strives to be abundant, and no one should be penalized for successful cultivation when it is not under the control of the State or corporations. Individuals frequently raise safety concerns to justify limits. However, when questioned about what is unsafe, the response is that individuals may sell the materials. It was not related to any direct safety issues or toxicity of the plants and mushrooms. In this view, safety concerns were for corporate interests' bottom line and a strategy to suppress competition. DN's approach is to remove all limitations on plant and fungal material that grows in the ground, allowing local community culture and trade to flourish.

Another policy red flag is the closed task force. The closed task forces formed in Oregon and Colorado comprise individuals from a primarily punitive worldview, with most members consisting of law enforcement, legal specialists, and addiction professionals. There needs to be more feedback from the community or those who understand how to participate in these experiences. Creating a closed task force allows corruption and corporate influence to affect laws and regulations. Already members of Oregon's closed task force had conflicts of interest, compelling members of the rules board to resign. Will this be a repeat in Colorado? The closed task force should take measures to ensure that feedback from the public is sought out and respected in the process. DN calls for open task forces in which people from various communities may share their perspectives and contribute to developing protocols that focus on healing rather than profiteering.

The current legalization frameworks are also causing data collection and privacy concerns. Gathering data for behavior analysis and product sales is also known as "surveillance capitalism." Shoshana Zuboff, professor at Harvard Business School and author of *The Age of Surveillance Capitalism*, defines it as "the unilateral claiming of private human experience as free raw material for translation into behavioral data. These data are then computed and packaged as prediction products and sold into behavioral futures markets—business customers with a commercial interest in knowing what we will do now, soon, and later." In other words, surveillance capitalism collects data to predict a person's behavior and market more products to them. Mason Marks has already raised the alarm in Oregon and Colorado, where the client's data is in danger. Although Oregon's M109 initially stated that their data would be private and confidential, Marks points out that their new strategy is to create a loophole to access the client's data; mandating that they give their consent to sharing to use the healing center's services. However, this becomes even more concerning when discussing transformative or profound healing experiences that might render the person hyper-suggestible or vulnerable. Imagine a record of data being purchased and sold by independent parties who have examined the private aspects of a person's psychedelic experience. Will they attempt to increase product consumption, influence the person's worldview, or relate this data to a social credit score? Public health research is essential, but it should not be done at the expense of anonymity or with a consumerist agenda. Data is the currency of the modern era; demand protection right away.

The legalization agenda also places these sacred medicines in the hands of the government and bureaucracy. The laws and regulations in Oregon and Colorado have a two-year deliberation and implementation phase. However, the constraints imposed by closed task forces gradually erode the rights of individuals and what they voted on. For example, in Oregon, due to a need for uniformity, only one variety of mushroom, Psilocybe cubensis, will be permitted, excluding several beneficial mushroom species. Participation in these regulated systems is costly and excludes indigenous and spiritual practitioners unless they pay thousands of dollars for a state-issued therapeutic license. The consumer will often shoulder these exorbitant expenditures and estimates of $3000–5000 for luxury retreat facilities are common. Furthermore, others are concerned about using a state-run treatment modality for these experiences, preferring culturally relevant venues outside of state-run institutions. We are already witnessing the impacts of bureaucracy in Oregon, where it has taken over two years to implement the project. There are even reports of their funding running out shortly, requiring Oregon taxpayers to foot the rest of the bill. The implementation and service expenses will keep these therapies out of reach for the poorest and most marginalized people. DN envisions laws and regulations being developed from the ground up. In other words, we advocate for providing safeguards for individuals who are currently practicing so that they can participate in therapeutic work without obtaining an expensive state-approved license when no study suggests that this training will be any more beneficial than what is already available.

Grassroots Approach

The worldwide push to emphasize clinical and medical settings is inaccessible in terms of both expense and ethos for individuals in need, with therapy costing thousands of dollars to acquire material that exists naturally in entheogens. Personal sovereignty and the power to determine our human experience are fundamental human rights. Adults are not children. With the proper knowledge, we can empower individuals to make their own decisions, which is a powerful first step toward healing. We do not have to look for solutions in a dysfunctional system. We can generate fresh ideas. We should rethink our approach to these plants and mushrooms by stressing a personal and community-use strategy. We are not seeking to develop the next antidepressant as a band-aid to societal issues without addressing the underlying causes of mental health in society. We recognize that entheogens may profoundly alter consciousness, and we must create a new culture to accommodate those changes. We can establish an infrastructure that protects and encourages the responsible use of these entheogens, enabling us to learn from them and each other.

A community-based approach to entheogens has several advantages. To begin with, culturally relevant settings and information will have a higher chance of reaching individuals needing healing who may not be able to pay for thousand-dollar therapies. We envision a healer-led approach that honors community voices already engaged in and rooted in these practices. Being a part of the community makes it more difficult for corporate influence to co-opt or buy out these services. Trust, compassion, and integrity are critical values that will be more likely to be adopted when profit is not the primary motivation. We propose a local economy by decriminalizing the "grow-gather-gift" model. This approach would enable a giving economy where individuals would pay for services, but plants and mushrooms remained complimentary. We need time to reestablish a culture that has been torn apart by years of criminality, shame, and persecution of peoples across the world. We must prioritize collective well-being, mutual respect, and shared prosperity to rebuild this broken trust.

It is essential to develop new and innovative ways to solve the problems we face. A few possibilities for how this may look necessitate humanity to think creatively. For example, these experiences could take place in a collective where people pay to join, and there are possibilities for cultivation, guidance, and community integration. Alternatively, we could establish sacred gardens and nature-based libraries, emphasizing teaching while providing access to plants and mushrooms for personal home grows. We could encourage knowledge and resource sharing by leaving a cutting or spore and taking a cutting or spore. We could also create a seed exchange and bartering system that would offer people opportunities to grow their gardens. Whatever the future holds, continuing to decrease stigma and educate the community about the ability to produce and share their entheogens will assist the most marginalized members of society.

There is good cause for optimism and confidence; we are winning the war for agency over consciousness. A growing narrative supporting people's empowerment and liberation is shattering the veil of medicalization and corporatization. There is legitimate worry about the dangers of mainstreaming through medicalizing and legalizing policies. Nevertheless, the decriminalization movement is already well underway. All that remains is to continue educating the public and reforming policies to provide equal access for all. We can always ensure that individuals may engage in these spiritual and sacred experiences by using community-based approaches to access. Psychedelic healing is unlike the cannabis market, where people buy cannabis products regularly. A significant transformative entheogenic experience may occur just a few times each year, if at all. The anticipated billion-dollar market is likely overstated. Who needs to purchase from corporate interests when we can saturate the market with home growers and community practitioners? So, stand strong, honor the progress that is being made at the grassroots level, and keep fighting for the unalienable right to have a relationship with nature. Let us reclaim our ancestral sacred earth medicines and ensure that healing can continue in our communities.

9

The Future of Psychedelic Research

Interview with Rick Strassman, MD

As an undergraduate, Dr. Rick Strassman majored in zoology at Pomona College in Claremont, California, for two years before transferring to Stanford University, where he graduated with departmental honors in biological sciences in 1973. He attended the Albert Einstein College of Medicine of Yeshiva University in the Bronx, New York, where he obtained his medical degree with honors in 1977.

Dr. Strassman took his internship and general psychiatry residency at the University of California, Davis, Medical Center in Sacramento, and received the Sandoz Award for outstanding graduating resident in 1981. From 1982–1983, he obtained fellowship training in clinical psychopharmacology research at the University of California, San Diego's Veterans Administration Medical Center. He then served on the clinical faculty in the department of psychiatry at UC Davis Medical Center, before taking a full-time academic position in the department of psychiatry at the University of New Mexico School of Medicine in Albuquerque in 1984.

At UNM, Dr. Strassman performed clinical research investigating the function of the pineal hormone melatonin in which his research group documented the first known role of melatonin in humans. He also began the first new US government approved clinical research with psychedelic drugs in over 20 years, focusing on DMT, and to a lesser extent, psilocybin.

Dr. Strassman's DMT: The Spirit Molecule, *an account of his DMT and psilocybin studies, has sold a quarter-million copies as of mid-2021, and been translated into over a dozen languages, including Mandarin. He co-produced an independent documentary by the same name, which was the most-streamed independent drug documentary on Netflix. He also is the author of* DMT and the Soul of Prophecy, Joseph Levy Escapes Death, *and a co-author of* Inner Paths to Outer Space. *He currently is Clinical Associate Professor of Psychiatry at the University of New Mexico School of Medicine and lives in Gallup, New Mexico.*

Interview conducted by Natalie L. Dyer, PhD

∞

ND: How do you think psychedelic research can help encourage responsible use of psychedelics at a personal level?

RS: It should emphasize the importance of set and setting and characterize the optimal set and setting for whatever intent one goes into a psychedelic state.

ND: How can psychedelic research positively impact society, policy, and law?

RS: We don't yet have good models for taking psychedelics most productively. It may be possible to address and problem-solve certain societal problems using the effects of psychedelics, steered toward that purpose, in policymakers.

Policy regarding psychedelics might be better informed once the full range of experiences available through the psychedelic experiences is mapped out.

Legal issues are complicated. I think medical utility and safety under medical supervision are now established. But they are still highly abusable. I think a new drug Schedule ought to be established, say IA, where specially trained, certified, supervised, and regularly re-certified individuals can administer these drugs. Dropping these drugs to Schedule II or III would be catastrophic in my opinion, because anyone with a Schedule II or III license (veterinarians, dentists, any physician) could prescribe them.

ND: How can psychedelic research benefit humanity as a whole?

RS: There are a lot of ways these drugs might help society: therapy, creativity, spirituality, consciousness. But let us not gild the lily: there are plenty of ways they may be misused by the unscrupulous.

ND: What was the most challenging aspect of conducting psychedelic research, and how did you overcome it?

RS: There was lack of communication between various agencies necessary for human research with Schedule I drugs to take place. FDA and DEA needed to align themselves and that took a lot of ferrying back and forth.

ND: What were the various factors that motivated you to start doing psychedelic research?

RS: I was interested in the biology of spiritual experience. I saw many similarities between descriptions of the psychedelic drug state and those brought on by meditation methods. I thought there must be some common underlying biological processes taking place, to the extent the syndromes resembled each other.

ND: Please describe your hallucinogen rating scale. Why did you create it?

RS: It's a self-administered rating scale. I interviewed about 20 people who'd taken DMT in the non-research setting in order to get a sense of what to expect in my study. I categorized expected effects using both a psychiatric mental status approach as well one informed by the Buddhist psychological model laid out by the Abhidharma literature. Things like somatic, emotional, volitional, cognitive and perceptual effects. I wanted to quantify the DMT effect by having volunteers score items "not at all" to "extremely" for items that fell into these various categories.

The scale has been used around the world in various translations for many different drugs and about 50 peer-reviewed papers have published data derived from it.

ND: Can you explain the importance of set and setting in psychedelic use and how they impact research?

RS: Set is one's mental, physical state, expectations, memories, intent. Setting is the outside environment in which the drug is taken, including the set of those around you.

If you have never taken a psychedelic drug before, and are afraid of what might happen, taking a drug called psychotomimetic by the researchers, in a frightening research environment, supervised by people who don't like you, who think psychedelics are diabolic will lead to a different experience than if you've taken psychedelics before, had good experiences, are in a pleasant environment, with friends.

This will affect recreational use, as well as the findings of research studies. Early alcohol-abuse treatment studies didn't control for these factors and this is probably why results varied so widely. Newer studies are taking these factors into account and thus results are much more uniformly positive.

ND: Please briefly describe DMT. Did any of your participants discover God/Higher Power with DMT? If so, were any of them atheist or agnostic before this experience?

RS: DMT is related to serotonin and melatonin, made primarily in the lungs, also in pineal, probably retina.

One of the volunteers had a T-shirt made after his high dose experience that said: "There are no atheists at 0.4" which was our high dose. By and large, most people's religious sensibilities were validated or strengthened, but essentially unchanged, although one Catholic lost his faith after his big DMT session.

ND: How do psychedelics have the power to deepen spirituality?

RS: Psychedelics enhance the ability to apprehend usually invisible forces or processes. Thus, they may make apprehensible certain previously invisible things which might be called spiritual: insights, feelings, ideas, visions. They may also make people more narcissistic, walled-off, or insane. I don't see psychedelics as inherently spiritual; it depends how one approaches the experience, understands, and integrates it. These are functions of the intellect or rational aspect of the mind, which I don't think are nearly as affected as the more imaginative functions where feelings, perceptions are apprehended.

ND: Have you had personal psychedelic experiences? If so, can you describe one that had a deep transformative impact on you?

RS: I don't answer that. If I say I have, I'm accused of zealotry. If I say I haven't, I'm accused of not knowing what I'm talking about. But I have seen the power of psychedelics to impact people, for good and/or ill.

ND: If you could summarize the most important findings of your psychedelic research in one sentence, what would it be?

RS: Psychedelics enhance the imagination without a corresponding effect on the intellect.

ND: What advice do you have for new researchers hoping to get into the field of psychedelic research?

RS: Don't specialize too early. Study and take classes in areas that you believe might be enhanced by the application of the psychedelic experience; and that may enhance our understanding of the psychedelic experience.

Find a good psychotherapist with his/her feet on the ground, preferably a psychoanalytically-trained one. Enter into and dig deep into a spiritual tradition that you can relate to. The longer the better. Humble yourself. Continuously.

Don't think that people who do psychedelic research are any more enlightened than anyone else. That includes you.

10

Integrating Psychedelics and Western Medicine

Interview with Rick Doblin of MAPS

Rick Doblin, PhD, is the founder and executive director of the Multidisciplinary Association for Psychedelic Studies (MAPS). MAPS is a high-profile nonprofit research and educational organization that develops medical, legal, and cultural contexts for people to benefit from the careful uses of psychedelics and marijuana.

Doblin received his doctorate in Public Policy from Harvard's Kennedy School of Government, where he wrote his dissertation on the regulation of the medical uses of psychedelics and marijuana and his Master's Thesis on a survey of oncologists about smoked marijuana vs. the oral THC pill in nausea control for cancer patients. His undergraduate at New College of Florida was a 25-year follow-up to the classic Good Friday Experiment, which evaluated the potential of psychedelic drugs to catalyze religious experiences. He also conducted a thirty-four year follow-up study to Timothy Leary's Concord Prison Experiment.

Rick studied with Dr. Stanislav Grof and was among the first to be certified as a Holotropic Breathwork practitioner. His professional goal is to help develop legal contexts for the beneficial uses of psychedelics and marijuana, primarily as prescription medicines but also for personal growth for otherwise healthy people, and eventually to become a legally licensed psychedelic therapist. He founded MAPS in 1986, and currently resides in Boston with his wife and puppy, with three empty rooms from his children who have all graduated college and begun their life journeys.

Interview conducted by Ocean Malandra

∞

OM: Hi Rick, thanks for taking the time today. Let's start at the beginning. Can you tell me why you started MAPS back in 1986? What was the need that you felt needed to be addressed at that time?

RD: The reasons for starting MAPS in 1986 were basically the same reasons that I decided back in 1972 to focus my life on psychedelics. At this time I was undergoing psychedelic therapy myself, and knew I wanted to play a role in bringing back psychedelic research.

I think psychedelics have a fundamental role to play in human survival, human thriving and cultural transformation. What I think was clear to me in '72, before I even learned about MDMA, was that the problems of human survival were not going to be solved by technology. While technology can help us to feed more people, help us to use more resources, the key problems were psychological.

We need people to be coming from this new globalized mystical spiritual state, where people realize we are all connected on a very fundamental and important deep level. Not just with each other, but with the animals, with the earth, and the universe. If people had that fundamental sense of mystical unity, then they would be less prejudiced, less willing to be irrational, less willing to demonize and scapegoat others.

Then we would find ways to work together with other people. And the alternative is what we see now with this rise in tribalism and fundamentalism, the exclusion of the "other." What we see in America is a turning away from our basic American values, away from welcoming immigration as a nation of immigrants. So, it just seemed to me that psychedelics have the potential to produce these mystical experiences that have fundamental political implications.

Recent studies from neuroscientists all over the world have shown that psychedelics like MDMA can catalyze compassion and empathy. We need to understand how compassion and empathy are generated and how we generate more of it. MDMA can also help break cycles of emotional trauma. We carry hatreds and conflicts from one generation to the next. Now I'm not saying psychedelics are the only way to mystical experiences or the only way to heal trauma, or the only way to promote empathy and compassion, but I think for most people, they are more reliable than 20 years of meditation, or mindfulness training or other things.

So it was really the political implications that drove me to devote my life to psychedelics in '72 and then to create MAPS in '86.

OM: OK, so since you brought up MDMA and seem to be focused on it right now I want to ask about that particular psychedelic. The FDA recently granted "breakthrough" therapy designation to MDMA for the treatment of PTSD right? This greenlighted you guys to go ahead with Phase III trials with MDMA, which would be human trials, and the final trials right?

RD: Yes, and we were actually granted a Phase III protocol back in 2016. And so we could have started with those Phase III trials right away, but we decided to enter this special protocol assessment process with the FDA where we basically work with them on every step of the process. So, it took until 2018 to develop this protocol where the FDA agreed with every element of our Phase III trial design. That is what we really needed, the "breakthrough" therapy designation was like a bonus, it's like a gold star. It's the most important program the FDA has to facilitate the most promising drugs. We were not sure they would give that to us because it's very political, it's very public, and if they were worried about backlash from politicians, they need to be very careful about giving that to us.

But the fact that the FDA gave us that "breakthrough" designation is the confirmation that the FDA is putting science before politics. So now we have many more meetings with the FDA, they have a shorter time to reply to our queries, and we are starting to talk about commercialization issues. They have indicated that this is a very promising "breakthrough" therapy and they have gone all out to try to help us.

So ever since 1992, when the FDA had some advisory committee meetings that put science before politics when it comes to Schedule 1 research, they really have stayed true to their word. And granting us "breakthrough" therapy designation was the ultimate demonstration of that.

OM: Cool, so it sounds like they have accepted the fact that there is therapeutic potential in MDMA right?

RD: They have totally accepted that. And here is a way to help you understand that. A couple of years ago we had a meeting with the FDA to discuss our marijuana and PTSD study. We are also doing a study with 76 veterans who are using marijuana to treat their PTSD. And the study uses four different kinds of marijuana. One with 9–12% THC, with virtually no CBD, one with 12% CBD and virtually no THC, one with about 8% of both THC and CBD and then a placebo as well.

So, after this teleconference with the FDA, and we had already been given approval for this study, I told them, "I just want you all to know, our priority is MDMA, because with MDMA-assisted therapy in just a few sessions, surrounded by psychotherapy, the opportunity is there for people to be cured of PTSD." Where with marijuana, marijuana reduces the symptoms of PTSD, and people can feel a lot better, but their problems tend to come back when they don't take the marijuana.

I said that to them. And then the response I got was, "Rick, you don't have to apologize for just treating symptoms, in psychiatry, that's about all we do." So, I think that when they see that MDMA does offer the opportunity for a cure, I mean you don't cure psychosis, you don't cure schizophrenia usually. Occasionally they will cure depression and anxiety, but they mostly just treat symptoms. So, I think they recognize that there is something remarkable here, and that is what the "breakthrough" for MDMA-assisted psychotherapy for PTSD is all about. It does offer an opportunity for a cure.

OM: That kind of leads into my next question. So, is there a describable mechanism for how psychedelics are able to do this? Is it because of a deeper introspection or a particular property of psychedelics? What do you think it is about psychedelics that can catalyze this true healing, not just treatment of the symptoms?

RD: Well, I think there is what is called "neuroplasticity." This means that psychedelics cannot just change your perceptions in the short run, but also what happens is a change in your brain's organization in the long run. What I mean by that, and particularly with MDMA for PTSD, is that so there is a reduction of activity in the amygdala, the fear processing point of the brain, and there is an enhanced connectivity between the amygdala and the hippocampus, which is where memories are stored. And there is an enhanced activity in the frontal cortex, where we put things in context.

There is also a release of oxytocin and prolactin, which is associated with promoting bonding among nursing mothers and lovers. It's kind of like the love hormone. It builds connection and therapeutic alliance. Plus, there are changes in the serotonin and dopamine neurotransmitter levels. But what happens is, people with PTSD have, in a sense, a memory that has not been fully processed. It's sort of stuck in the short-term memory and doesn't get processed into a long-term memory. It gets replayed in dreams, it gets replayed in the day when things trigger it, causing people to mute their experiences with emotional numbing.

Once you have got PTSD it just keeps going and going and going and you are never really free of it. Under MDMA, with the reduction of fear and the enhanced connectivity with the hippocampus, people are able to process the memory and then store it in the frontal cortex as a more long-term memory. So that the way in which your brain processes information about things that related to the trauma has been changed. What we see from the MRI studies is that with people with PTSD who have been successfully treated, there is less activity in the amygdala than before.

But let me just add one thing. From the point of view of the FDA, and from the point of view of what we are trying to do, which is to make MDMA into a medicine, we have to prove both the safety and the efficacy of the drug. But you don't have to have a mechanism of action. You don't have to understand how it works. So, we are not focusing on that, but, as more and more people try to figure out how MDMA works we are getting more and more information. We now have people doing studies where PTSD patients under the influence of MDMA will be scanned by MRI as the process takes place.

I think what we are going to see is that the processing of emotionally challenging content is happening in a different way than normal. And that processing, well let's put it like this. Let's say you have a river running downstream and then you have this dam blocking it up. And the dam is the fear, the fear that is blocking the processing. But if you dig this channel around the dam then you are able to process, and before you know it the water is running in a new way. And I think that is what is going on metaphorically with memories.

OM: Correct me if I am wrong, but I was reading some of the MAPS bulletins and it appears that low doses can even increase fear and that it takes a full dose, a full-blown psychedelic experience, for this therapeutic effect to occur right?

RD: Yeah well, what we show is that people get a little better with therapy alone, even with a placebo. But when you add MDMA you get twice as much benefit or more. However, with low dose MDMA, when you add that to therapy, people do still see some benefit but not as much as if you had given them an inactive placebo. So, what's happening is that people with PTSD react so much to this fear of their fear-based memories that it makes them unable to process the trauma.

When they are in therapy, and they are there in order to try to focus on the trauma, and we give them low dose MDMA, it activates them and it activates their memories, but the fear is not reduced. So consequently, people can feel worse, they can feel uncomfortable, and sometimes want to drop out. I thought low dose MDMA would be the ideal solution to the double-blind problem, but it actually just causes more confusion. And that's why for the Phase III trials the FDA has said use inactive placebo, do not use low dose MDMA.

OM: Interesting. Before I let you go, I do want to ask you about plant medicines like Peyote, Iboga, Ayahuasca and Psilocybin Mushrooms, most of which have also been shown to be promising treatments for PTSD as well as other disorders like depression, anxiety, addiction and more. Is it going to be harder to integrate those plant medicines into the modern health field?

RD: That's a very, very interesting question. So, there is this whole concern about cultural appropriation. These plant medicines come with their own traditions and so how do we blend that with modern medicine? So, there was actually a discussion at the International Transpersonal Conference in Prague that I attended recently, and someone's talk was about the "shadow" side of psychedelic research. The issue was this cultural appropriation.

And as I was listening to her talk I was thinking, "I'm so glad MDMA was just invented in 1912," and also LSD was invented, so there can be no appropriation. But my view of the issue is this, there is a lot of indigenous wisdom that has been building up for millennium for various spiritual purposes. But they also have a lot of cultural and religious dogma wrapped up into them. The União do Vegetal for example is patriarchal, it's homophobic, it's been influenced by the Catholic Church. So I feel fundamentally that these drugs are the inheritance of the entire world, not just the group that discovered them.

I feel that it's possible and ethical to take ayahuasca and take it out of religious contexts and do research with it. Look what we have done with psilocybin. Psilocybin came from the Mazatec Indians and now we are doing synthetic psilocybin for all kinds of people with all kinds of problems. I think we need to honor where these drugs come from, but we are not imprisoned by them or required to use them in one particular way. Things have to evolve over time and in different culture contexts.

When we are talking about therapy though, it is a little bit hard to get standardized doses from these plants. But in many cases I think the plants are better than the isolated compounds by themselves. Certainly, marijuana for example is better in plant form than isolated cannabinoids. On the other hand, some cannabinoids, like CBG for example, which has been shown to possibly shrink tumor cells, you need to isolate to get enough of an effect. Which you will never get just by smoking marijuana.

I think that the plants will find a way into psychotherapy. Ayahuasca, for example, can be standardized sufficiently, and so I do think that this will happen. But I also think that there is a romantic aspect to this, which is wrong, which is that if it's from nature it's good and if it's from the laboratory it's bad. I don't buy that. I think that we have had incredible results with synthetic psilocybin that, you know, didn't come from a mushroom, it was created synthetically.

OM: And so let's talk about the overall picture then. And feel free to add your own experiences. But what role do you think psychedelics have, and you touched on this in the beginning, in helping us create a more harmonious society for humanity here on earth?

RD: I don't mean to say it like this but I think that if we don't legitimize psychedelics, I'm not sure we are going to survive as a human race. By legitimize I mean mainstream and make them legal not just medically but also accessible for personal growth, for deepening spirituality, for a deepened connection with nature, for romance and couple's therapy. Those things are not medical conditions but for example we can never medicalize MDMA for couple's therapy, because that's not a disorder. You can't medicalize psychedelics for lack of a mystical experience. Because that's not, at least presently, considered a medical condition.

The destruction of nature is happening at an increasing rate. The dangerous rate at which more and more people are getting nuclear weapons is also frightening. I do think there are lots of ways out of this other than psychedelics, to bring about this sort of mystical connection which has these attendant political implications. And again, it's not just the chemicals, but the combination of the psychedelic experience with therapy and other conditions that makes the change. But I do think that given the lack of time that psychedelics have the potential, I think they will play a major role in global transformation.

11

The Promise of Classic Psychedelic-assisted Treatment for Criminal Recidivism: A Call to Action

Sara N. Lappan, PhD & Peter S. Hendricks, PhD

Sara N. Lappan, PhD, MFT completed her doctorate in the department of Human Development and Family Studies with a concentration in Couple and Family Therapy at Michigan State University. She is interested in health behavior and its effects on the family unit. More specifically, she is interested in creating/adapting family programs aimed at reducing rates of obesity as well as a systems-based application of psilocybin for survivors of sexual assault and rape.

Peter S. Hendricks, PhD is Professor and Director of Research in the Department of Health Behavior of the School of Public Health at the University of Alabama at Birmingham (UAB). His research centers on the development of novel and potentially more effective treatments for substance dependence, with specific areas of focus on cigarette, cocaine, cannabis, opiate, and polysubstance dependence in vulnerable populations (e.g., individuals in the criminal justice system).

∞

In the United States crime is a major societal and public health issue. Indeed, there are over 6 million adults under criminal justice supervision in the United States, amounting to about one in every 38 adults, or 2.6% of the entire adult population (Kaeble & Cowhig, 2018). This is the highest reported incarceration rate in the world. The two most common criminal offenses comprise drug-related crimes (possession, use, distribution, or manufacturing of illicit substances), which account for 14% of all arrests, and property crimes (burglary, larceny/theft, motor vehicle theft, or arson), which account for 13.5% of all arrests (FBI Uniform Crime Report, 2015). Though violent crime (murder, rape/sexual assault, assault, or robbery) accounts for only 5% of all arrests, it is among the most costly to society. Of note, the lion's share of crime in the United States is directly caused or significantly exacerbated by substance use (Chandler et al., 2009).

The costs of criminal behavior are vast and include medical expenses, lost earnings, property damage and loss, psychological distress and decreased quality of life, as well as expenses related to law enforcement and correctional programs (McCollister et al., 2010). The prevention of recidivism is, therefore, a significant priority. However, recidivism in the criminal justice system is alarmingly high with about two-thirds of released prisoners being arrested for a new crime within three years and more than three-fourths being arrested within five years of release (Durose et al., 2014). Furthermore, though several interventions have been developed to decrease recidivism, these interventions demonstrate limited effectiveness (Ferguson & Wormith, 2013; Pearson et al., 2002; Visher et al., 2005). There is thus an urgent need to explore alternative approaches to prevent criminal behavior.

The notion that classic psychedelics might prove useful therapeutic agents in the reduction of criminal behavior occurred to a small handful of researchers during the first wave of classic psychedelic research in the mid-twentieth century. Tenenbaum (1961) provided ten treatment-resistant sex offenders with multiple sessions of LSD-assisted group therapy and noted meaningful therapeutic gains in all but one of his patients. Arendsen-Hein (1963) administered several LSD-assisted therapy sessions to 21 "criminal psychopaths" and reported similar therapeutic effects in 14 of his subjects. Timothy Leary (1969) attempted to use classic psychedelics to reduce recidivism in his Concord Prison Experiment, but methodological shortcomings preclude any definitive conclusions (Doblin, 1998). Sadly, despite these promising initial findings, misinformation, stigma, lack of funding, and legal proscription of classic psychedelics ruled out further study.

Of course, research with classic psychedelics is currently experiencing a modest yet growing renaissance, and since 2014 three studies have investigated the relationship between classic psychedelic use and criminal behavior. Hendricks et al. (2014) evaluated data collected from 2002 to 2007 on over 25,000 substance-involved individuals charged with a felony and under community corrections supervision in the Deep South of the United States. They found that use of hallucinogens, a broader class of substances characterized primarily by classic psychedelics, was prospectively associated with a reduced likelihood of recidivism. Use of virtually every other class of substance, by contrast, was associated with an increased likelihood of recidivism.

Walsh et al. (2016) examined the prospective associations between naturalistic hallucinogen use and intimate partner violence (IPV) among 302 inmates at a United States county jail. Analyses revealed that any lifetime use of hallucinogens was associated with lower rates of IPV; approximately 27% of the hallucinogen-use group was arrested for later IPV compared with approximately 43% of the group that reported no hallucinogen use.

Additionally, the relatively few participants who met criteria for a lifetime hallucinogen use disorder were less than half as likely as those without such a disorder to be arrested for later IPV. Finally, Hendricks et al. (2018) tested the relationships of classic psychedelic use with criminal behavior among over 480,000 United States adult respondents pooled from years 2002 to 2014 of the National Survey on Drug Use and Health while controlling for a number of potential confounding variables.

The authors found that lifetime classic psychedelic use was associated with a reduced likelihood of past year larceny/theft, past year assault, past year arrest for a property crime, and past year arrest for a violent crime. Similar to previous results, lifetime illicit use of other substances was, by comparison, largely associated with an increased likelihood of these outcomes.

In sum, contemporary findings, coupled with older results, suggest that interventions making use of classic psychedelics could very well be effective in reducing recidivism. Limitations to the extant research, however, are obvious. The older studies lack control conditions and rely on subjective clinical impressions as opposed to standardized measures or objective recidivism outcomes, and the newer studies are correlational in nature, which leaves open the possibility that characteristics of classic psychedelic users, rather than classic psychedelic use itself, may be responsible for the results.

Only placebo-controlled, randomized clinical trials can definitively determine whether classic psychedelics might reduce criminal behavior. In the meantime, an important question is why or how classic psychedelics might prevent recidivism. The answer, as recently proposed by Hendricks (in press), may lie in the discrete emotion known as *awe*.

As explained by Hendricks (in press), awe is experienced whenever one encounters a stimulus so vast and novel— immense and outside of one's current understanding—that one must change their view of reality. Perhaps most importantly, awe promotes the *small self*, which includes feelings of oneness with others, reduced individualistic tendencies, and devaluation of self-relevant goals. Awe is believed to promote social integration and cooperation, which are evolutionary advantageous behaviors very much key to humanity's success. Classic psychedelics, Hendricks (in press) suggests, may ultimately occasion profound awe.

Thus, for those struggling with substance use, including those who account for the preponderance of arrests made in the United States, the experience of discrepancy between a hedonic pursuit and a cause greater than self (e.g., family, community, or a belief system) may provide the impetus for prolonged sobriety. For those engaging in antisocial transgressive behavior, an experience that emphasizes the interdependence of all life may foster greater concern for others, and consequently, reduced criminality. These effects may be reflected in changes to the default mode network (DMN), a network of brain regions that appear to be hyperactive and hyperconnected among those with various mental health conditions (Carhart-Harris et al., 2012; Whitfield-Gabrieli & Ford, 2012).

Obviously, these hypotheses require further testing, and the question remains whether classic psychedelic use has a direct, clear, and causal impact on recidivism and decreased criminal behavior. Given the staggering costs of crime and disappointing outcomes associated with existing recidivism treatments, this question should be addressed urgently. While political hurdles associated with the completion of a clinical trial of classic psychedelic-assisted treatment of criminal behavior certainly exist, exacerbated perhaps by the association of classic psychedelics with the countercultural revolution of the 1960s, these obstacles can be mitigated with rigorous, professionally conducted research. The reactionary backlash to the tumult of the 1960s spawned both the War on Drugs and the subsequent expansion of the carceral state, which are now two major causal factors of both excessive incarceration and recidivism rates (Conyers, 2012).

Some might say it would then be poetic, or perhaps a bit of cosmic humor, that these same substances, once the subject of moral panic, could play a role in the solution to mass incarceration, which was in no small part spurred by that very moral panic. Nevertheless, the current objective is to revolutionize forensic mental health care for the better, not to initiate a cultural revolution. The nation has grown weary of the War on Drugs and its unending litany of unintended consequences. Cultural attitudes have become more permissive regarding harm reduction as opposed to punitive approaches; most Americans now live in a jurisdiction with at least some access to medical cannabis, and even politicians are embracing public health models of addiction policy (Pew Research Center, 2014). Indeed, the time is ripe for change.

We ask only that policymakers go where the evidence leads, and point to the success of recent investigations suggesting that the classic psychedelic psilocybin may promote abstinence from tobacco (Johnson et al., 2014; 2017) and alcohol (Bogenschutz et al., 2015), and elicit enduring prosocial behavior among healthy non-institutionalized participants (Griffiths et al., 2018). Moreover, we emphasize the importance of administering classic psychedelics only under the most carefully controlled conditions using all appropriate medical, psychological, and ethical safeguards. It is imperative that individuals in the criminal justice system be allowed to make an informed and voluntary choice to participate in any trial involving the administration of a classic psychedelic.

We also emphasize that reasons for breaking the law are varied, and these reasons should be considered when designing a trial of classic psychedelic-assisted treatment. Those who commit crimes as a matter of desperation or survival in a demanding and impoverished criminogenic environment may be less likely to respond to classic psychedelic treatment as compared to those who are suffering from addiction or demonstrate psychopathic traits (e.g., a fundamental disregard for others and/or a lack of empathy; Hare, 1996). It is possible that the positive feedback loops reinforcing psychopathic traits (Preszler et al., 2018) could be interrupted by a classic psychedelic-occasioned mystical experience. This is a question for future research. Nevertheless, we submit here that criminal behavior born of psychopathic traits might be the most prudent target of classic psychedelic interventions in this arena. Of course, substance use disorders are obvious (and ongoing) targets as well.

In conclusion, classic psychedelics may represent viable tools in the prevention of criminal recidivism, with promising preliminary findings and a theoretical model explaining a potential mechanism of action. However, whether classic psychedelic use might directly reduce criminal behavior is an open question. We encourage the conscientious investigation of this important question so as to address a crucial societal and public health problem.

References

Arendsen-Hein, G.W. (1963). LSD in the treatment of criminal psychopaths. In: R.W. Crocket, R.A. Sandison, & A. Walk (eds.). *Hallucinogenic Drugs and Their Psychotherapeutic Use* (pp. 101–106). London: H.K. Lewis & Co. Ltd.

Bogenschutz, M.P., Forcehimes, A.A., Pommy, J.A. et al. (2015). Psilocybin-assisted treatment for alcohol dependence:

A proof-of-concept study. *Journal of Psychopharmacology*, 29, 289–299.

Carhart-Harris, R.L., Erritzoe, D., Williams, T. et al. (2012). Neural correlates of the psychedelic state as determined by fMRI studies with psilocybin. *PNAS*, 109, 2138–2143.

Chandler, R.K., Fletcher, B.W., & Volkow, N.D. (2009). Treating drug abuse and addiction in the criminal justice system: Improving public health and safety. *JAMA*, 301, 183–190.

Conyers, J. Jr. (2012). The incarceration explosion. *Yale Law & Policy Review*, 31, 377–387. Available at: http://digitalcommons.law.yale.edu/ylpr/vol31/iss2/4

Doblin, R. (1998). Dr. Leary's Concord Prison experiment: A 34-year follow-up study. *Journal of Psychoactive Drugs*, 30, 419–426.

Durose, M.R., Cooper, A.D., & Snyder, H.N. (2014). *Recidivism of Prisoners Released in 30 States in 2005: Patterns from 2005 to 2010* (NCJ 244205). Bureau of Justice Statistics Special Report. Washington, DC: United States Department of Justice, Federal Bureau of Investigation, Criminal Justice Information Services Division. Available at: https://bjs.ojp.gov/content/pub/pdf/rprts05p0510.pdf

FBI Uniform Crime Report (2015). *Crime in the United States 2015*. Washington, DC: United States Department of Justice, Federal Bureau of Investigation, Criminal Justice Information Services Division. Available at: https://ucr.fbi.gov/crime-in-the-u.s/2015/crime-in-the-u.s.-2015

Ferguson, L.M., & Wormith, J.S. (2013). A meta-analysis of moral reconation therapy. *International Journal of Offender Therapy and Comparative Criminology*, 57, 1076–1106.

Griffiths, R.R., Johnson, M.W., Richards, W.A. et al. (2018). Psilocybin-occasioned mystical-type experience in combination with meditation and other spiritual practices produces enduring positive changes in psychological functioning and in trait measures of prosocial attitudes and behaviors. *Journal of Psychopharmacology*, 32, 49–69.

Hare, R.D. (1996). Psychopathy: A clinical construct whose time has come. *Criminal Justice and Behavior*, 23, 25–54.

Hendricks, P.S. (in press). Awe: A putative mechanism underlying the effects of classic psychedelic-assisted psychotherapy. *International Review of Psychiatry*.

Hendricks, P.S., Clark, C.B., Johnson, M.W. et al. (2014). Hallucinogen use predicts reduced recidivism among substance-involved offenders under community corrections supervision. *Journal of Psychopharmacology*, 28, 62–66.

Hendricks, P.S., Crawford, M.S., Cropsey, K.L. et al. (2018). The relationships of classic psychedelic use with criminal behavior in the United States adult population. *Journal of Psychopharmacology*, 32, 37–48.

Johnson, M.W., Garcia-Romeu, A., Cosimano, M.P. et al. (2014). Pilot study of the $5\text{-HT}_{2A}R$ agonist psilocybin in the treatment of tobacco addiction. *Journal of Psychopharmacology*, 28, 983–992.

Johnson, M.W., Garcia-Romeu, A., & Griffiths, R.R. (2017). Long-term follow-up of psilocybin-facilitated smoking cessation. *The American Journal of Drug and Alcohol Abuse*, 43, 55–60.

Kaeble, D., & Cowhig, M. (2018). *Correctional Populations in the United States, 2016*. Washington, DC: United States Department of Justice, Office of Justice Programs, Bureau of Justice Statistics. Available at: https://www.bjs.gov/content/pub/pdf/cpus16.pdf

Leary, T. (1969). The effects of consciousness-expanding drugs on prisoner rehabilitation. *Psychedelic Review*, 10, 29–45.

McCollister, K.E., French, M.T., & Fang, H. (2010). The cost of crime to society: New crime-specific estimates for policy and program evaluation. *Drug and Alcohol Dependence*, 108, 98–109.

Pearson, F.S., Lipton, D.S., Cleland, C.M. et al. (2002). The effects of behavioral/cognitive-behavioral programs on recidivism. *Crime & Delinquency*, 48, 476–496.

Pew Research Center (2014). America's New Drug Policy Landscape. Available at: http://www.people-press. org/2014/04/02/americas-new-drug-policy-landscape/

Preszler, J., Marcus, D.K., Edens, J.F., & McDermott, B.E. (2018). Network analysis of psychopathy in forensic patients. *Journal of Abnormal Psychology*, 127, 171–182.

Tenenbaum, B. (1961). Group therapy with LSD-25. (A preliminary report). *Diseases of the Nervous System*, 22, 459–462.

Visher, C.A., Winterfield, L., & Coggeshall, M.B. (2005). Ex-offender employment programs and recidivism: A meta-analysis. *Journal of Experimental Criminology*, 1, 295–316.

Walsh, Z., Hendricks, P.S., Smith, S. et al. (2016). Hallucinogen use and intimate partner violence: Prospective evidence consistent with protective effects among men with histories of problematic substance use. *Journal of Psychopharmacology*, 30, 601–607.

Whitfield-Gabrieli, S., & Ford, J.M. (2012). Default mode network activity and connectivity in psychopathology. *Annual Review of Clinical Psychology*, 8, 49–76.

12

On the Frontlines: Legal Advocacy, Drug Policy and Psychoactive Plants

Constanza Sánchez Avilés & Andrea Langlois, ICEERS

Constanza Sánchez Avilés *is the Law, Policy & Human Rights Director at the International Center for Ethnobotanical Education, Research, and Service (ICEERS). She is a political scientist and holds a PhD in International Law and International Relations. Her main areas of work and research are national and international drug control policies and the intersection between drug control, human rights, and social justice.*

Andrea Langlois *is the Director of Engagement for ICEERS. She is a communicator, facilitator, and community-based researcher and is passionate about policy, dialogue, and social movements. She holds a Master's Degree in Media Studies from Concordia University in Montreal, Canada, and spent several years engaging in and writing about alternative media. She is the co-editor of two books on autonomous media and pirate radio.*

The International Center for Ethnobotanical Education, Research, and Service (ICEERS) *is a nonprofit organization dedicated to transforming society's relationship with psychoactive plants. They do this by engaging with some of the fundamental issues resulting from the globalization of ayahuasca, iboga, and other ethnobotanicals.*

∞

Introduction

Practices involving psychoactive plants and fungi—such as ayahuasca, iboga, San Pedro, psilocybin mushrooms and coca leaf, among others—are becoming more globalized, traveling beyond their traditional territories and into communities on all five continents. While it is difficult to estimate the types or number of people offering or partaking in these experiences or integrating them into their ways of life, there are many signs pointing to the extent of this expansion. For example, there are increasing numbers of groups with online and social media presence[1] and both mainstream and alternative media outlets have taken notice.[2]

We also know that religious groups that use these plants in their rituals, for example the Santo Daime and the União do Vegetal ayahuasca churches, now have numerous chapters in countries around the globe and their memberships are growing.[3] There is also a growing number of researchers working to cut through political red tape and raising funds independent of traditional funding sources to study the effects, risks, benefits of these ethnobotanicals and looking into their potential for addressing some of society's most challenging health issues (such as depression and addiction) at their roots.[4]

Although these practices have a long history of use in many cultures and regions, their globalization is relatively recent. While the benefits that these practices bring to individuals and communities is notable, challenges are also arising as once localized, integrated practices make their way into a global context marked by the interpretation of psychoactive plants and concoctions as "drugs," thereby bringing them into the sphere of government regulation. One of the unintended consequences of the migration of these ethnobotanicals, rituals, and practices into "non-native" countries is that they are increasingly becoming tangled in the net of international and national drug control mechanisms.[5]

The severity of this issue was brought to the attention of the International Center for Ethnobotanical Education Research and Service (ICEERS), then a relatively new non-governmental organization, in January 2010, when we were asked for help by two individuals whose ayahuasca ceremony in Chile was raided by police. They were charged with drug trafficking and endangering public health, and in addition to the legal charges, they suffered persecution within the media, which profoundly affected their personal and professional lives. ICEERS accompanied them on their journey to fight for their right to follow their calling to offer healing opportunities to their community. A solid human-rights focused defense strategy with the documents provided by the defense, which included correspondence from the INCB that confirmed that ayahuasca is not subject to international control, and the testimonies by scientific experts and of ayahuasca session participants in the sessions contributed not only to the acquittal of the accused, but also to the recognition by the judges—in a pioneering way—that participants had benefited from their experiences with ayahuasca. Also notable, is that this 2012 ruling stated that ayahuasca was not included in the prohibition of DMT.[6]

The Manto Wasi case was the first of many to come, and laid the foundation for supporting others facing criminalization for their practices with psychoactive plants, such as, but not only, ayahuasca. In 2016 this work became formalized as the Ayahuasca Defense Fund, a program within ICEERS that has the goal of creating a context within which traditional psychoactive plants can be used legally and safely. Over the course of our first official year, the ADF team engaged with over 40 in-depth legal queries and we also worked with 20 criminal cases in 11 countries. Since we embarked on this journey, there have been many lessons learned about what happens when traditionally-used psychoactive plants meet the criminal justice and drug control systems. This chapter is a short overview of some of our experience with litigation and provides an overview of trends that we have observed along the way with regards legal prosecution and policy development around psychoactive plants.

Tradition Meets the Drug Control Regime

Traditional psychoactive plants have much longer, culturally-rich histories than the contemporary drug control system. Although tobacco, coffee, and cacao were widely embraced by colonial interests and became important commodities since the expansion of European colonialism, psychoactive plants used in ceremonial settings did not find their ways onto the merchant ships of empires. When the international drug control system began to take shape at the beginning of the twentieth century, opium, cannabis and coca leaf were the plants onto which controls were imposed. The 1961 United Nations Single Convention on Narcotic Drugs, which unified all the previous treaties on this subject, did not include plants such as *Banisteriopsis caapi*, *Psychotria viridis*, mescaline cactus, psilocybin mushrooms, mild stimulant plants like khat or ephedra, or other plants such as kava kava, kratom or *Salvia divinorum*. However, dimethyltryptamine (DMT), mescaline, psilocybin and other psychoactive compounds that are present in traditional plant-based materials were scheduled under the UN Convention on Psychotropic Substances of 1971. Although these "active" components are scheduled, the International Narcotics Control Board (INCB) has clarified that none of these plants, nor concoctions that contain them, currently fall under international control.[7] In practice, however, this has been differently interpreted by policymakers and judges in individual countries.

Although only a few countries, such as France,[8] have explicitly prohibited some of these ethnobotanical materials, in most states they exist in a legal limbo, which means that those who import, organize ceremonies and use these ethnobotanicals are exposed to significant legal uncertainty. Based on our observations (although no rigorous data has been gathered), there have been more arrests and prosecutions in many parts of the world for activities related to these practices since the 90s, and particularly after the INCB drew attention to this question in its 2010 Annual Report.[9]

These legal incidents represent a clash between cultures—the result of what happens when ancient rituals and knowledge encounter a modern drug control regime based on the values of social control, sobriety, and that do not distinguish among different types of substances or different types of uses. In the view of these authorities, any use equals abuse, and the only legitimate uses for mind-altering substances are medical and scientific ones. As traditional practices have traveled around the globe, it has not only been the ingestion of plants or concoctions that have been transported, but most people (particularly in the case of ayahuasca and iboga) engage in practices that are guided, and that incorporate ritual or ceremony, and increasingly a focus on "after-care" and integration. Ritual and ceremony are essential components of traditional use, which is embedded within a cultural context and the safety net of tradition, which are elements that continue to be held as important as these practices are adapted into new contexts. In these new contexts, once they come to the attention of governments, psychoactive plants like ayahuasca are framed as dangerous and are stigmatized as a social and public health problem. Taken out of their cultural contexts, they are problematized and the solution offered is that of prohibition, criminalization, and the prosecution of individuals and groups. This happens, in part, because traditional psychoactive plant practices do not fit into the traditional frameworks on which drug policies are enacted in modern societies. These types of practices that use psychoactive plants do not fit into the traditional dichotomies of medical and scientific uses versus recreational uses, and the settings are not exactly traditional indigenous, nor therapeutic, nor recreational.[10]

From Rituals to Molecules: Drug Policy Reductionism vis-à-vis Psychoactive Plants

Modern drug control systems do not take culture into account (at least not *all* cultures), or the fact that there are beneficial uses for psychoactive plants. Therefore, when someone is arrested for possession of the plants themselves, such as coca leaf powder or dried mushrooms, or a concoction like the ayahuasca brew, or when Churches request the state for permission to use these ethnobotanicals in their religious rituals, the authorities are left needing to "fit" them into existing categories and structures. There are several trends that we are seeing with regards to this phenomenon.

Using ayahuasca as an example, the first trend is that drug control bodies are increasingly classifying it as a "novel psychoactive substance" (NPS).[11] Despite ayahuasca's long history of use in the Amazon, awareness of the brew did not exist widely in the 1960s when many other psychedelics rose to popularity,[12] which meant that it also did not face demonization or was included in the repression of psychedelics at that time. It is only in the last 10–20 years that the brew has begun to literally travel the globe, as ayahuasca-drinking practices of diverse types have become more common in many countries. As ayahuasca and other traditional plants come onto the radar of drug control bodies and monitoring agencies, they have frequently been labeled as NPS—a category typically used to describe research chemicals, legal highs or spice (synthetic cannabinoids) designed to mimic existing established recreational drugs with no history of human use or scientific data about their effects and health risks.

The second trend is that the legal interpretation applied to traditional plants is extremely reductionist and complex cultural psychoactive plant practices are generally reduced to being characterized not just as the psychoactive plant preparation, but merely as some of their active components. Ayahuasca becomes DMT, San Pedro cactus becomes mescaline, and so on. This approach by the authorities neglects the substantial scientific evidence available about their safety, and divorces the plants from their cultural, historical, religious, ritual, and social backgrounds. This has been the case even when indigenous people have been arrested in European or North American countries. As countries grapple to find a legal "place" for these ethnobotanicals, they are interpreting laws, or passing new ones, to regulate ayahuasca and other traditionally-used plants more strictly.

As noted above, while the INCB has acknowledged the increasing popularity of traditionally-used psychoactive plants and have written that they are not under international control (even though the chemical equivalents of alkaloids they may be found to contain are controlled under the 1971 Convention on Psychotropic Substances), they also warn governments about their alleged health risks, without citing any evidence, and invite member States to control them at the national level. For example, Italy scheduled iboga in 2016[13] following a legal case involving a contaminated sample and Canada has recently added ibogaine to the prescription drug list.[14] In 2016, the UK passed a controversial and problematic Psychoactive Substances Act, aiming to make "any substance that stimulates or suppresses the central nervous system," and is not regulated, illegal by default.[15] Similarly, Hungary passed new regulation for new psychoactive substances in 2012 that may end up impacting traditional plants.[16]

The last trend we have identified is that there are many religious groups working to obtain legal recognition and permission to use psychoactive plants in their rituals—especially religious groups that use ayahuasca. In this area, there have been both successes and setbacks. People hoping to seek legal recognition for their spiritual practices are looking to the historic case of the União do Vegetal (UDV) in the USA[17] and the recognition of ayahuasca churches in Brazil for inspiration, and we are seeing an increase in ayahuasca, Santo Daime, and UDV churches that are seeking exemptions or other forms of legal recognition from the state, particularly in Canada, the USA, and Denmark.

The results of these procedures have been uneven. While in Canada,[18] the Ceu do Montreal (a Santo Daime church) and the UDV obtained permission to import ayahuasca for ritual use in 2017, in the Netherlands the news was less encouraging. At the beginning of 2018 the Court of Amsterdam decided that import, possession and sacramental use of ayahuasca within the Ceu da Santa Maria (a Santo Daime church) was not permitted anymore as it "leads to an unacceptable danger to public health."[19] While there has been some progress, it's also clear that the landscape for religious groups seeking legal permissions for ayahuasca use is not predictable at this time.

Conclusions

The increasing legal prosecution of psychoactive plant practices has been one of the unintended consequences of their globalization—although the exact scope is unclear since there is no data about the expansion of use. Since 2010—and more systematically since the establishment of the Ayahuasca Defense Fund (which not only covers ayahuasca but also other traditionally-used plants)—ICEERS has worked to research and follow legal and policy developments related to psychoactive ethnobotanicals so that we can support the community in legal harm reduction, promote the protection of human and indigenous rights, and advocate for progressive policy change.

Due to the scope of our work, and how states are responding to and framing psychoactive plants in countries where practices do not have historical roots, our efforts have been interwoven with drug policy reform attempts and focused on supporting individuals facing criminalization.

Drug policy is fluid—it is not set in stone; even when it appears to be, there are most certainly cracks within the system. Recent advances in drug policy reform—for example the decriminalization of cannabis in several parts of the world, the accommodation of religious ayahuasca practices in some jurisdictions and the growing support for psychedelic therapy—show promise for the future of these plants within international and domestic drug policies. While this represents an opportunity to achieve legal status for psychoactive plants at the national level, efforts are required to monitor how drug policies are being applied so that we can continue to advocate for drug policies based on public health, human rights, scientific evidence, human development and civil society participation.

As a community we need to take steps in stewarding the expansion of practices with psychoactive plants and contributing to social change that result in drug policies and social structures that acknowledge their value for the betterment of society. And, within this stewardship, it is essential to offer support and solidarity to Indigenous peoples for and the recognition of their inalienable right to continue their long traditions of ceremonial use of sacred plants.

References

1. While there is no general directory of ayahuasca retreat centers, groups, or churches, there are several sites, such as PsychedelicExperience.net, that seek to be portals, and at the time of writing a basic search within Facebook for ayahuasca groups resulted in nearly 100 groups listed.

2. Media from the *Guardian*, to *Men's Health*, and *VICE* have published pieces about ayahuasca, and ibogaine has recently garnered additional attention because of the opioid overdose crisis in the US and Canada.

3. See Bia Labate, The Santo Daime and the UDV receive religious exemption to use ayahuasca in Canada, available at http://www.bialabate.net/news/the-santo-daime-and-the-udv-receive-religious-exemption-to-use-ayahuasca-in-canada (June 2017).

4. This topic has been recently discussed at the 2018 Commission on Narcotic Drugs meeting in Vienna: The right to science and freedom of research with scheduled substances, available at: https://www.fuoriluogo.it/oltrelacarta/video/the-right-to-science-and-freedom-of-research-with-scheduled-substances/#.WsSw_WYrxTb (March 2018).

5. For more data on arrest and legal incidents worldwide, please see the Ayahuasca Defense Fund Annual Report for 2016–2017, available at: https://www.iceers.org/wp-content/uploads/2020/06/ICEERS_ADF_annual-report-2016-2017-FINAL.pdf. The authors would like to acknowledge Bia Labate for her invaluable contribution to this data gathering and analysis.

6. Constanza Sánchez & José Carlos Bouso, Ayahuasca: de la Amazonía a la aldea global, Drug Policy Report 43, Transnational Institute (December 2015). Available at: https://www.tni.org/files/publication-downloads/dpb_43_spanish_web_19122015.pdf

7. See INCB, *Report of the International Narcotics Control Board for 2012*, Paragraphs 329–330 (p. 46), available at: http://www.incb.org/documents/Publications/AnnualReports/AR2012/AR_2012_E.pdf

8. In 2005 France scheduled *"Banisteriopsis caapi, Peganum harmala, Psychotria viridis, Diplopterys cabrerana, Mimosa*

hostilis, Banisteriopsis rusbyana, harmine, harmaline, tétrahydroharmine (THH), harmol, harmalol." (See: Arrêté du 20 avril 2005 modifiant l'arrêté du 22 février 1990 fixant la liste des substances classées comme stupéfiants.) This happened right after the Supreme Court confirmed the acquittal of several members of a Santo Daime church.

9. See INCB, *Report of the International Narcotics Control Board for 2010,* Paragraphs 285–287. Available at: http://www. incb.org/documents/Publications/AnnualReports/AR2010/ AR_2010_English.pdf. In Paragraph 287, the Board states: "that, in view of the health risks associated with the abuse of such plant material, some Governments have placed certain types of plant material and preparations under national control. The Board recommends that Governments that have not yet done so and have experienced problems with regard to persons engaging in the recreational use of or trafficking in such plant material, to remain vigilant (since the risks associated with such use may increase) and to notify the Board and the WHO of those problems. The Board recommends that Governments should consider controlling such plant material at the national level where necessary."

10. See Sánchez & Bouso, 2015.

11. This has been the case, for example, in Spain, where several traditional plants are included in the "early warning system" of the national drug control strategy. Plant materials such as ayahuasca, kratom or kava kava are often seized in Spanish customs. See official reports at: http://www.pnsd.msssi.gob. es/profesionales/sistemasAlerta/home.htm

12. Tupper, K.W. (2008). The globalization of ayahuasca: Harm reduction or benefit maximization? *International Journal of Drug Policy,* 19(4), 297–303.

13. Italian update of psychotropic substance table, August 1, 2016: http://www.gazzettaufficiale.it/eli/id/2016/08/11/16A05919/sg%3Bjsessionid=q2AypY7DnATPsx1NyXfeiA__ntc-as4-guri2b

14. May 19, 2017 notice from the Government of Canada: https://www.canada.ca/en/health-canada/services/drugs-health-products/drug-products/prescription-drug-list/notice-prescription-drug-list-multiple-additions-2.html

15. Literally, Article 2 of the Act states: "For the purposes of this Act a substance produces a psychoactive effect in a person if, by stimulating or depressing the person's central nervous system, it affects the person's mental functioning or emotional state; and references to a substance's psychoactive effects are to be read accordingly." See full text at: http://www.legislation.gov.uk/ukpga/2016/2/contents/enacted

16. See Péter Sárosi, Drug Law in Hungary—Drugreporter's Guide for Foreigners (2013). Available at: https://drogriporter.hu/en/drug-law-in-hungary-hclus-guide-for-foreigners/

17. The UDV has an extensive online archive about the case, see: http://udvusa.org/supreme-court-case/

18. Rochester, Jessica. How Our Santo Daime Church Received Religious Exemption to Use Ayahuasca in Canada (July 17, 2017). Available at: http://chacruna.net/how-ayahuasca-church-received-religious-exemption-canada/

19. More details on the case at Rini Hartman, Dutch Freedom of Religion on Pause for Santo Daime (March 2018). Available at http://chacruna.net/dutch-freedom-religion-pause-santo-daime/

13

Psychotropic Resistance in the Amazon Rainforest

Riccardo Vitale, PhD, Anthropologist

Riccardo Vitale is an Italian self-defined "liberation anthropologist." Which means bringing knowledge to the communities rather than extracting from them, and building knowledge with the communities for the communities. He obtained a PhD from Cambridge University with a thesis about the Zapatista movement in Chiapas, Mexico. His expertise covers human rights, armed conflict, social movements, indigenous politics, gender relations within social movements, sustainable development, resilience, climate change adaptations and indigenous practices of yagé medicine, spirituality, and resistance.

Riccardo is a former adviser of a plethora of international humanitarian and development agencies: Oxfam America, the UNHCR, the Norwegian Refugee Council, ICG and GIZ, amongst others. Since 2016 Riccardo works as a full-time adviser for the Union of Indigenous Yagé Medics of the Colombian Amazon (UMIYAC). He is also a member of the ayahuasca tech team of the Indigenous Medicine Conservation Fund (IMC), an indigenous led fund.

∞

The spiritual forces that inhabit the earth are manifesting themselves in the dreams and visions of traditional healers and through warnings, which we name "natural disasters." These cataclysms are the last warnings of Mother Earth calling for a global change of conscience that leads to the emergence of sustainable and organic relations between humans, all other species, and the planet.[1]

In 2016, while working with Oxfam America, I had the opportunity to research how indigenous Nasa communities in Colombia coped with and responded to a series of volcanic eruptions and avalanches that could have greatly disrupted their lives. The research showed that indigenous discourses and adaptive practices present viable solutions to pressing issues such as disaster risk reduction, climate change, and sustainable rural development. The study concluded that the assimilation of indigenous knowledge into the development agenda is a necessary precondition to the achievement of the 17 Sustainable Development Goals (SDGs) adopted by the nations of the world in the 2030 Agenda for Sustainable Development.[2] A precondition to a friendlier, healthier, more peaceful and livable planet for all species, to put it simply.

There has never been a better time for dialogue, as indigenous knowledge and scientific thought have never been so close. The Anthropocene is defined by *Homo sapiens'* activities as prime drivers of major changes in the earth's system, human induced transformations that are so substantial they require their own geological era. The Stockholm Resilience Centre, the leading research hub specialized in ecology, accepts Anthropocene as a useful geological category and warns that humans are about to surpass planetary boundaries, beyond which global ecological sustainability will become impossible. The message being conveyed is that the contemporary system of production, consumption, distribution, and fuels burning is compromising the existence of all species on planet earth. In other words, earth's regenerative capacities are being stretched to the limit and we are about to surpass several tipping points.

This is also what indigenous Latin American movements are contending from the front lines of an extractive economy. There is no such thing as inert matter; everything in the cosmos is alive and all things (beings) are interconnected. The dominant modes of production and consumption are antithetic to the mission of preserving and caring for mother earth. A mission consigned to indigenous peoples of the Americas by the Law of Origin, Fundamental Law and Customary Law. These ancient laws and precepts are transmitted from one generation to the next as part of the body of oral knowledge, norms and beliefs that form the cosmovisions (worldviews) of first nations tracing their origins to pre-Columbian civilizations.

In this essay, I will share some observations from my ongoing experience as full-time advisor for a grassroots organization called the Union of Indigenous Yagé Doctors of the Colombian Amazon (UMIYAC). Yagé is the generic term used in Colombia for the "psychoactive" mixture known worldwide as ayahuasca. This botanical brew is a primary element in the spiritual/medicinal knowledge systems of at least 160 indigenous peoples from the Amazon basins.

In the following pages, I will talk about the relationship between yagé rituals and the ongoing struggle for the cultural revitalization and defense of the Amazon Rainforest vis-à-vis extractive businesses such as cattle ranching, mining, hydrocarbon exploitation and narcotrafficking. I will touch upon the importance of spiritual medicine in peace building at the community level, where it most matters. Specifically, the role played by UMIYAC's doctors as healers of personal and collective traumas; as promoters of reconciliation; and as agents for the reconstruction of communities whose social fabrics have been torn apart by decades of war. The chapter includes an account of the problematic relationship between indigenous communities and international, conservationist NGOs, and how these interactions affect local autonomy and self-determination, key concepts for indigenous peoples in Colombia and beyond. A second account relates to an aspect of so-called new age markets and the psychedelic renaissance, and how the incipient "ayahuasca tourism" in the Amazonian foothills is bringing in some useful cash, but also interfering with communities' spiritual practices.

International NGOs

Note: although I wanted to disclose the name of the International NGO, I am taking the advice of close friends and allies and have decided not to do so. I have already received a threatening letter on behalf of this INGO from an uncool law firm specialized in assisting big oil and other extractive corporations, including one accused of "disappearing" environmental activists. Any reader who really wants to find out the name can do so easily enough. However, the intention of this piece is not to exacerbate old beefs, but rather to amplify local voices and remind stakeholders operating in conflicts zones to always abide to Do No Harm policies (which means, avoiding creating divisions at all costs).

The UMIYAC covers communities in the Departments of Putumayo, Caquetá, together with some binational villages on both sides of the Colombian-Ecuadorian fluvial borders. These regions are a quagmire of conflicting interests, and communities are struggling to come out of a fifty-year-long war profoundly linked to issues of land tenure, cocaine production, hydrocarbon exploitation, cattle ranching and mining. This forgotten conflict has so far produced almost seven million internally displaced people (IDPs) and over nine million registered victims of war.[3]

The UMIYAC's members belong to five Amazonian ethnic nations: Siona, Inga, Cofán, Koreguaje and Kamëntšá. The organization's mission consists of protecting indigenous territories and the rainforest; preserving ancestral medical knowledge systems and practices; and strengthening indigenous autonomy, governance, and self-determination.

In 2016, I was summoned by a council of elders and asked to join the UMIYAC as an advisor. The organization was experiencing a profound crisis and some of the members thought that incorporating new professionals would be a good step towards strengthening UMIYAC. Whilst excited by the opportunity, I was wary of the complexities of indigenous politics, and I approached the new responsibility with some caution. Not even three months into my new collaborative role, we got news that UMIYAC's sole sponsor, a known conservation INGO, decided to cut us off. They disagreed with the new board of directors, questioned the fact that they were incorporating advisors not hired by the INGO, and strongly criticized the election of a new president whom they could not control.

To keep afloat, the UMIYAC would have to learn how to raise funds, a task for which nobody had any previous experience. With no budget to hire a professional fundraiser, the new scenario put the future of the UMIYAC at risk. Yet, aside for a couple of exceptions, the reaction amongst the members was one of relief and renewed enthusiasm. Not a surprise, considering that several people had told me that the international INGO they had been working with was bossy, disrespectful of local cultures, and out of tune with realities on the ground. It didn't end there.

The new UMIYAC team had prioritized revitalizing the alliance between male traditional healers and their female counterparts grouped in another local and grassroots organization of botanical practitioners and midwives, all belonging to the same communities. To prevent this alliance, the INGO used its political and economic influence to keep the two organizations as separate as possible. This went as far as actively provoking a fracture between the two organizations which, eight years down the line (I'm writing this in 2023), is still dividing the communities and the work of both organizations.

This is a serious breach of Do No Harm policies which mandate that international humanitarian agencies and NGOs do everything in their power to avoid causing divisions. Not to mention that the Colombian indigenous movement revolves around four fundamental guiding principles: autonomy, land, culture, and unity. But there is more. Within the framework of transitional justice (the jurisprudence designed to help society come out of the armed conflict), 34 of the 102 existing Colombian indigenous ethnicities have been declared at risk of "physical and cultural extermination" by the Constitutional Court (Resolution 004 of 2009). Following this landmark legislation, as part of a mandatory governmental rescue initiative, each indigenous people at risk compiled a Safety Plan, in which they identified all threats to their survival. Quite worryingly, on page 111 of the Safety Plan redacted by the Cofán people, the international NGO in question (together with another gigantic wildlife fund) is mentioned as a driver of cultural and physical extermination for violating the right to previous informed and free consultation (PIFC) and weakening unity, grassroots organizing and territorial sovereignty.

Out of Sight, Out of Mind

Piedra Alegre rests in a luscious valley nestled between two rivers in the mountainous region of the Colombian Amazon rainforest.[4] In this part of the world, green manifests itself in all the hues perceivable by the human eye. This village of about three hundred families is part of a large indigenous territory in the Department of Putumayo, Colombia. *Resguardo* (reservation) is the term used by Colombian jurisprudence to define lands legally assigned to indigenous people. They are immune to seizure, inalienable and imprescriptible.[5]

In the 1990s, Putumayo was awash with coca plantations— and still is to a large extent. Local economy revolved around the cocaine and the hydrocarbon (exploration and extraction of crude oil) industries. At these latitudes, coca-paste production is the most rational choice a peasant family can make. Once processed, the cocaine alkaloid enters a market generating annual revenues of 300 to 500 billion dollars a year, most of which is skillfully laundered and injected into the world's financial markets and banking systems. Cocaine production's annual yields surpass the GDP of rich countries such as Sweden and Belgium.[6] Although they are stuck at the least profitable stage of the production and supply chain, Colombian farmers know that no other local produce can compete with coca cultivation. People from Piedra Alegre, like most rural dwellers in their region, started relying on coca cultivation since the early 90s.

Often, epicenters of production of hyper-globalized commodities, such as cocaine, gold and oil, are portrayed as far away, remote, marginal and forgotten "no-man-lands." Perhaps this is a clever mechanism to create "faraway alterities" intended to protect the consciences of people in the global north and avoid inconvenient scrutiny. The incongruity of this POV becomes evident if one tries to visualize the intricate networks of routes, physical and digital, over which money, and the powdery merchandise transit at any given time. The Amazonian foothills, Cartagena, Bogotá, Geneva and Abu Dhabi, Houston, Dallas and Dubai, Chicago, Mexico City, New York, Los Angeles, London, Paris and Rome, Gioia Tauro and Moscow, Frankfurt, Antwerp, Lagos, Monrovia, Cape Town and Sydney, to name but a few, are all connecting banking and seaport nodes in the transactional map of the cocaine economy.

In the year 2000, Bill Clinton's Plan Colombia landed in Putumayo, with its thousands of tons of Monsanto's glyphosate, helicopters, aircraft, and pricey military contractors. This, together with a sharp increase in paramilitary violence, set the stage for massive internal displacement. Thousands of families sought refuge in the neighboring state of Nariño, bringing the coca economy with them. In 2012, I was carrying out interviews in Tumaco, Nariño. The work was part of a large study by Oxfam America about perceptions of international aid in conflict zones (Cohen, Vitale 2013).[7] At the time, Tumaco, Colombia was experiencing a severe humanitarian crisis: fighting amongst rival armed groups, forced displacement, violence, extreme poverty. As it turned out, the people I interviewed, aid workers, community activists, NGO personnel, government officials, army and police, all attributed the crisis to the relentless fumigation and policing operations that hit neighboring Putumayo as part of Plan Colombia. As coca cultivation moved to Nariño, the Plan Colombia machinery followed suit, producing similar if not worse results: increased violence, mass displacement, poisoning of the environment. Seeing this firsthand and hearing the testimonies of so many people from all walks of life convinced me that the war on drugs was and still is a lucrative trade. And that draconian drug policies in Colombia and abroad have nothing to do with ending cocaine production and consumption; in fact they seem to exist in total symbiosis with narcotrafficking.[8]

People in Piedra Alegre learned from experience what spraying massive quantities of glyphosate can do to an environment. The herbicide, that used to be marketed by Monsanto as Roundup, contaminated their lands and waters, killing their subsistence food crops. Alberto, a friend of mine who is a young healer and member of UMIYAC, told me:

Seeing how fumigation poisoned our environment and our lives helped us think about the effects of all chemical herbicides. To some of us, this devastation became vivid during yagé ceremonies when we would feel the pain we are inflicting upon Mother Nature and ourselves. The "easy" cash of the coca business blinded us; thank god, and thanks to our sacred yagé medicine, we relearned how to listen to mother nature and we heard the land screaming for mercy.

In 2010, the people of Piedra Alegre collectively chose to manually eradicate all coca fields. This move, negotiated with local municipal authorities, prompted the removal of Piedra Alegre from the list of areas targeted by aerial fumigation. Alberto said:

The coca economy never brings good things to a place. Money is wasted as fast as it is earned.[9] The raspachines (coca day laborers) spend all their free time in the cantinas getting drunk; prostitution is rampant; people carry firearms; they forget about community life and collective duties; violence increases; we lose all spiritual connections with the land, and once that happens, communities disappear.

Until the signing of the peace accords in 2016, the guerrillas were never far from the mountains surrounding Piedra Alegre. Coexisting with legal and illegal armed groups caused enormous stress on all villagers. As other rural communities throughout Colombia, the population of Piedra Alegre suffered from the utter disregard of International Humanitarian Law (IHL) by all parties involved in the conflict. On two occasions, hundreds of people in the village were caught in the crossfire. Ducked on the floor, terrified, bullets piercing their wooden houses. Everyone in the village has a war story to tell and everyone I know has lost at least one family member to the armed conflict. Such is life when you share your homeland with the most sought-after commodities in the history of humanity.

Spiritual Unity

The accepted understanding in the Amazon foothills is that the true owners of spiritual knowledge are elderly *curacas* (traditional doctors) of A'I, or Cofán, ethnicity. It is not infrequent for *curacas* to be invited to officiate ceremonies by other communities. Similarly, people from other regions travel south to attend ceremonies with A'I *curacas* in their territories along the natural fluvial borders separating Colombia and Ecuador. Furthermore, there are families from Piedra Alegre and other surrounding villages who also come from lineages of yagé doctors. The memory of their *tigrero* grandfathers and great-grandfathers still evoke reverence and admiration.[10] In Piedra Alegre, men, women and children as young as five years old partake in the *tomas de yagé* (yagé drinking sessions). The *tomas* takes place at night in appositely built ceremonial houses. These locations are called *casas de remedio* (healing houses) or malokas, a term borrowed from the Murui/Huitotos tribes.

One healing house in the vicinity of Piedra Alegre is particularly alive, with ceremonies taking place two or three times per week. *Curacas* who visit this healing house are often impressed by the numinous aura pervading the location, and the pristine and beautiful natural surroundings. Here, the roaring waters of the Caquetá River, the concerts of insects, amphibians, and mountain animals; the winds, the squalls and the thunders, together with a myriad of other indecipherable sounds of nature, offer a perennial and magnificent soundtrack. The house consists of a rounded wooden structure covered by an intricate, skillfully thatched roof, built according to traditional Ingano architecture. The house belongs to Celestino, a traditional doctor who has been practicing yagé medicine for over 20 years.

Like most other knowledgeable healers, he is a humble and understated person. In fact, my impression is that in the world of Amazonian spiritual medicine, the more you know, the less you say. Spiritual knowledge is a very sacred and private affair and is only shared in intimate, ceremonial spaces amongst elders and a few trusted apprentices. A sharp contrast from the burgeoning, highly competitive world of ayahuasca understood by writers, pharmacologists, academics, psychedelic advisors, new-age psychologists, and other experts.

My Inganos friends are passionate and talented musicians. So, harmonicas, flutes and drums are often present during the long ceremonial nights. Talking and storytelling are also important. Healing houses are privileged spaces for collective discussion and reflection. People talk about problems affecting their villages, hunting, fishing, or personal issues. Politics, the armed conflict, the hurdles of peasant life, injustices, and government's shortcomings are also common topics.

Poking fun and laughing are also common features in most yagé sessions. Jokes can be directed towards a fellow drinker, instigated by an anecdote or by something that a person has said or done. At times, jokes are wittingly carved around particularly sensitive issues. This turns injustices, loss of work, lack of money, and even violent episodes into life parodies that can trigger bouts of unstoppable, loud, and liberating collective laughter. Within UMIYAC, the Cofán stay true to their reputation of being the naughtiest and funniest *taitas*, always in the mood for jokes, teasing, and a good laugh. This behavior constructs complicities and inside-jokes that are repeated for years on end, during ceremonies and in everyday life. As with music, oral communication and storytelling, laughter and jokes are important complements in a person's curative process.

Transcending War Traumas and Resisting Hydrocarbons

From a public health perspective, post peace-accords Colombia shows alarming figures. Of the approximately 337,054 inhabitants of the Department of Putumayo, 43.7% have been registered as victims of the armed conflict. Considering that Putumayo is a rural department and that families are extended and consist of more than two people, this means that almost all the population of the department has had to live through a violent and traumatic event. The figures are equally alarming in the departments of Cauca and Caquetá, where respectively 23% and 44% of the total populations are registered as victims of the conflict (RUV, 2017). The rest of rural Colombia is, unfortunately, no exception.

In indigenous areas torn by the protracted war, access to health facilities can require long and expensive journeys. Communities, however, can treat a large plethora of conditions relying on the local pharmacopoeia and medicinal knowledge. In this context, yagé ceremonies are possibly the only spaces where people suffering from the effects of traumatic events can seek and receive attention. Yagé ceremonies, under the guidance of experienced healers, provide the specialized support needed by individuals and groups to process and transcend difficult and painful experiences. During ceremonies, people bond with one another in a spiritually heightened state of conscience. In this milieu, relationships amongst people from different communities and ethnicities are forged and kept alive. This process is vital when it comes to reconstructing or keeping together the social fabric of communities rendered highly vulnerable by centuries of colonial campaigns, ruthless religions, extractive economies, and the current fifty-year-long internal war.

This is why one of the UMIYAC's most interesting activities is a community health program called spiritual health brigades. The spiritual health brigades consist of one or more groups of traditional healers, accompanied by at least one elder, that move all year round from village to village. The locations are prioritized depending on the urgency of the requests received by the UMIYAC's office in Mocoa, Putumayo. The petitions are varied and can come from a group of families seeking cure for sick loved ones; from governing authorities needing to solve a collective problem, internal disputes or a safety issue related to external threats from narcotraffickers, or other organized armed actors menacing the villages. Rural bicultural schools and other indigenous organizations are always in the loop as they frequently request ceremonies to punctuate landmark events or important meetings.

The brigades end up fulfilling a plethora of purposes. They provide people living in war zones with constant mental and spiritual support; they give people living far away from health centers access to botanical medicine; they foster collective dialogue, helping communities strengthen unity and governance. And they provide a safe space for people living in forced confinement imposed by the armed groups to get together and share vital information about imminent threats and vulnerabilities that need to be addressed.

The long-term resistance of the Siona people of Buenavista against oil corporation Amerisur Resources PLC is an important example of how yagé medicine is helping communities cope in a complex and dangerous political arena such as is found in Colombia. By law, corporations interested in exploiting underground resources or intervening in any way within indigenous lands are obliged to engage in talks with local authorities. These mandatory negotiations are known as "previous consultations" and form part of the rights of indigenous people laid out in Convention 169 of the 1989 International Labor Organization (ILO), adopted in Colombia through Law 21 of 1991.

For years, the Sionas have had to cope with pressures from corporations wanting to explore and extract oil from their territories. As is the norm in these cases, the presence of the corporation and the negotiations caused confusion and quarrels amongst the villagers. Whilst a fundamental right, previous consultations often turn into long and debilitating processes, which can weaken the communities socially and politically. In 2010, I carried out a study for an international aid agency, which included understanding how Sikuani communities along the Orinoco River were coping with the presence of hydrocarbon companies. The material gathered for this work showed that nothing positive came out of repeated rounds of previous consultations. None of the development projects negotiated with Talisman Energy and other companies as paybacks for allowing seismic exploration ever translated into better living conditions. In fact, they provoked disagreement and rivalry amongst community leaders and spokespeople who accused each other of obtaining bribes and under the table compensations from the corporations.

At first, the Sionas fell into the same pattern as other indigenous communities and spiraled into divisions and internal feuds. This, however, changed when spiritual authorities (yagé healers) attached to different clans decided to get together and participate in the discussions through a series of yagé ceremonies. Through the spiritual practice, the communities consensually decided against oil exploitation of their territory. It was an important accomplishment because although indigenous communities cannot veto a hydrocarbon or mining project, having to face a united refusal and fearing bad PR and protracted negotiations, Amerisur temporarily opted to move to a different area. A small yet important victory.

Bonanzas

Yagé began leaving the jungle at the beginning of the last century through the emergence of syncretic spiritual practices propagated by followers of people such as Mestre Irineu in Brazil in the 1930s.[11] This migration gave rise to a movement of ayahuasca churches with wide international presence. Most renowned amongst these are: the Santo Daime, Barquinha and União do Vegetal (UDV). Since the 1990s, Colombian yagé healers or *taitas* from the Amazonian foothills started officiating ceremonies in urban settings. This trend consolidated over the years giving rise to disparate hybridizing urban modes of purveying and consuming yagé.[12] At present, yagé ceremonies are held in several countries every day, despite the blurry legal status of the brew. In fact, yagé has established itself as an international commodity for which demand keeps growing.

Within the UMIYAC, the global expansion of yagé is watched with ambivalence. Younger members who are also proficient travelers welcome the possibility to access growing markets in urban Colombia and abroad. At the same time, other associates are verbalizing concerns about burgeoning bad practices and about how psychedelic start-ups, retreats centers, and visionary entrepreneurs are already treating yagé as a new pharmaceutical bonanza.

A quick reminder. Quinine, rubber, hydrocarbons, gold, and cocaine, together with a long list of bioactive compounds, have historically generated huge fortunes as well as contributing to medical science and to the well-being of people everywhere. Bonanza was the term used to describe cycles of frenzied demand and massive extraction of Amazonian resources. Bonanza comes from the Latin word *bonus*, meaning good. But good for whom and at what price? The track record is bleak to say the least.

Two random references come to mind. The pages of José Rivera's *The Vortex*, and the scenes of *Embrace of the Serpent*, respectively a literary and a cinematic voyage through rubber plantations, enslaved natives, wandering amputees, burning villages, child raping priests, and sadists of all sorts. Complete the picture with the unsettling image of a despairing elder who lost his healing powers and can't figure out whether giving white people access to his sacred medicinal plants might help or worsen the fate of his already doomed tribe. Indeed, every time the Amazon rainforest becomes the go-to place for a hot resource, the price paid by indigenous communities is immense. Slavery. Rape. Massacres. Environmental devastation. Poisoning of the biosphere. Loss of territories. Mass displacements. Epistemicide. Genocide. "At risk of cultural and physical extermination" is how the Constitutional Court defined the predicament of 34 out of the 102 Colombian indigenous peoples (Writ 004, 2009). Five of these ethnicities at the brink of genocide live in the Colombian Amazon rainforest and use yagé for their medicinal and spiritual practices.

Psychedelic Markets

Psychedelic medicines are becoming globalized commodities, and it is common for some Colombian traditional healers to travel nationally and internationally, officiating ceremonies for burgeoning volumes of urbanites.[13] But from the POV of *taitas* and *seguidores*, the practice of *la medicina* remains inextricably enmeshed with its ecological and spiritual niche: the Amazon Rainforest.

The mastering of yagé knowledge (*el conocimiento*) is a lifetime endeavor that only particularly gifted individuals from selected lineages can withstand.[14] It requires physical and spiritual strength as well as years of complete dedication and commitment. Most *taitas* were initiated in their botanical and spiritual training at a very early age by their mothers, fathers, grandmothers, grandfathers or other relatives. Proper *tomas de yagé* should mainly take place in uncontaminated areas, in the plants' native ecosystem.[15] This is the ecological dimension where younger apprentices can learn from the elders and where yagé rituals achieve the highest transforming and curative efficacy. Moreover, elderly *taitas* and *curacas* are highly sensitive to environmental and spiritual pollution and must follow specific diets to preserve their spiritual strength and health. Frequent traveling to contaminated urban areas is often discouraged as it can affect the *taitas'* physical well-being, it can weaken their curative powers and it can compromise their spiritual knowledge. Who, where and how the brew is prepared will also influence the way participants experience a *toma*. Avoiding spiritual and ecological contamination of the places where the plants are grown is hence a crucial precaution. Yagé needs to live and grow in uncontaminated, bio-spiritual ecosystems accessed only by carefully selected cooks (*cocineros*). This is important, as the brew will absorb energies from those who prepare it. Anyone can potentially produce a psychoactive brew, but not just anyone can prepare a curative, spiritual medicine apt for a proper *toma de yagé*.

This is one of the reasons why the idea of any type of medicalized ayahuasca from the point of view of UMIYAC's healers sounds like an aberration and a marketing trick. In 2019, a young UMIYAC healer invited to an international ayahuasca conference in Catalonia was dismayed by the number of academics, anthropologists, psychologists, new age healers, writers and experts presenting at panels and giving talks about the wonders of this jungle medicine. *"I am a humble peasant healer,"* he said, *"my university is the forest, thirty years I've been practicing this medicine, and here I am, still at the beginning of my learning process and not an expert at anything, perhaps I've just been wasting my time."*

In the world of Amazonian yagé medicine, the more you know, the less you say. Your practice and reputation speak for you. Still, it is an obvious fact that traditional knowledge has become a commodity. Occasionally, visionary entrepreneurs and/or psychedelic start-ups approach the UMIYAC with business propositions. Sometimes they seek endorsement in exchange for monetary benefits; more than once they have asked for healers to become part of their board of directors. Professionals working for these organizations often mention reciprocity and shared benefits. The UMIYAC's position about the hyper-commercialization of traditional knowledge is unequivocal. Indigenous medicinal knowledge systems need to be nourished and protected.

Resilient botanical and spiritual cultural practices cannot be bought or sold. Reciprocity cannot be achieved when relationships are unequal and unbalanced. Communities at the brink of cultural and physical extermination because of cocaine, oil, gold, land and other extractive businesses must not be put in the position to negotiate for their survival. The Colombian constitution is one of the most advanced of the Americas and, along with international covenants such as ILO 169 and the 2007 UN Declaration on Indigenous rights, it provides judicial instruments of transitional justice, reparation and restorative justice that must be understood and applied whenever companies feel the need to approach indigenous communities regarding the use of local resources.

Spiritual Tourism

Since the signing of the peace accords, foreigners and Colombian urbanites have been transiting through areas of the country that used to be off limits. On a few occasions, whilst visiting the *malokas* of members of UMIYAC, I had the opportunity to partake in ceremonies where some of the participants were not patients or "yagé followers" from the region, as they are called in the communities. I am not talking about sporadic spiritual seekers or the friends that *taitas* make during work trips in Colombian cities, which is a fairly common occurrence. I am referring to relatively large groups of tourists led by a guide who have come to the region for the "authentic" yagé experience. These *tomas* turned out to be interesting and possibly alarming hybrids-in-process between local and commercial rituals.

On one particular ceremony day, I broke my own pre-ceremony rules by eating both a nice breakfast and a hefty lunch. Eggs, rice, lentils, veggies and quinoa soup, with fresh orange juice. An hour before going to the healing house, I also gulped a cup filled to the brim of a thick and delicious homemade *chicha*. That was after a tasty fish soup and yams. When in Rome...[16] Most *taitas* agree that entering a ceremony with a light stomach is good practice. However, people also say that having food a couple of hours before the ritual provides you with the necessary strength to endure a long night and intense yagé experience. The Cofán, for example, like to have dinner right before the ceremony. "You'll have something to puke," people often joke.

Walking towards the healing house, I got the familiar feeling of apprehension and expectation. We had a lot to talk about, and I was looking forward to being around the fire thinking, feeling and hopefully, yagé willing, discussing some pressing issues facing UMIYAC. The healing house was crowded, and upon entering, I noticed a group of about 20 foreigners and about ten locals. I also noticed that some of the habitual *drinkers* were not around. To avoid the intricate tangle of ropes, I chose to hang my hammock by the entrance. "Flirting with darkness, Riccardo?" said Celestino, after inviting me to hang my hammock next to his and that of other friends. Mistakenly, I paid no attention. There are no wasted words in indigenous spaces. How could I forget?

A few hours went by and I was having a hard time keeping myself together. The medicine had kicked in very strongly. The time I spent in the hammock surrounded by unknown bodies and absorbing undecipherable energies had also possibly taken its toll. Who knows. My head filled with obscenely violent images. There were no garments that could mitigate the bone-chilling cold which came from deep inside me. I felt like this was a gelid whisper from an eternal night in a parallel universe overpowering my every molecule and conscience. The only remedy I found that could melt this metaphysical ice was sitting by the bonfire and absorbing as much heat as possible. Fire is an important element of most yagé ceremonies. It purifies energies and stimulates thoughts; it is a communicator of wisdom and a protector from malignant beings. One of the most beautiful references about the relationship between humans and fire can be found in Akira Kurasawa's *Dersu Uzala*, when Dersu, an indigenous Nanai hunter, argues with the bonfire deep in the tundra, as Captain Arseniev watches in disbelief.

The bonfire always helps, so I made my way there and sat down on one of the makeshift wooden benches. My friends were already immersed in conversations. Land defense, social leaders, politics. Just a few weeks prior, one of UMIYAC's founders was killed as he was tending his land. His name was Mario Jacanamijoy, leader of the rural organization Tandachiridu Inganokuna. He was one of the 121 social leaders killed in Colombia in 2017.[17]

To me one of the most healing feelings of the yagé session is experiencing the flow of empathy and spiritual togetherness amongst people in the *maloka*. Maybe this is how pain gets processed, with the mediation of the medicine, the *taitas*, the followers, the chants, and the forest sheltering the whole experience. I don't know, what's certain is that this is how UMIYAC functions. Every event, every decision, every proposal and course of action gets processed, analyzed and discussed through yagé ceremonies.

My friends' voices were soothing and I listened attentively and participated in the discussion. Suddenly, I saw a shape leaning over me: *"This is an ayahuasca ceremony, can you two be quiet? We are here to concentrate!"* He says this to me in English. I'm now in a bit of shock. Should I pass on the message to the friend next to me? He is the person who left his house before dawn, chopped the vine, and collected the leaves for the herbal remedy we are all sharing. This is his family's healing house where he started drinking yagé as a kid. The words spoken during yagé sessions and around the fire are as important as the moments of silence. Customarily, people that cannot talk and need time alone do so in their hammocks.

Time for a cultural translation, suggested the naive and yagé drunken anthropologist in my head. So, I mumble a few phrases about land struggles and indigenous rights, trying to include him in the conversation. The tourist nods, but only to up the ante. Now he's telling me that he has done many, many ceremonies and that silence is important. Basically, he is adamant that we need to shut our mouths. I insist, "Violence is endemic in this area, and we are talking about..." He cuts me off! "Violence," he says, "is a karma that comes to those who attract it." I am now angry and offended. In fact, I found the comment about violence and karma hurtful and out of place, and I confess, it occurs to me that I could stick a burning log into this guy's eye.

Thank God I have too much respect for my friends' space, and it probably wouldn't be a good thing for the organization. How many times had I said and wrote that yagé ceremonies are also about peace building and reconciliation? Coherence, Riccardo. We are just passing egos bickering in the night. A friend whispered: *"He's playing with fire."* "Oh no!" I thought, "He's read my thoughts." There was tension all around. The tourist spoke again, but he was coming around because he switches to Spanish and cracks a joke. It's an anecdote about a scientist and a tiger in the jungle. The scientist is collecting samples and data nonstop as the tiger stays hidden and watches him. As the day ends, the scientist, now satisfied with his work, decides to make his way home. At this point, the tiger jumps out of its hideout, eats him up with one bite and exclaims: *El que sabe sabe.* The jokes is ambiguous and in Spanish it can have more than one meaning: "the one who knows, just knows" and "the one who knows tastes good"! Everyone burst into laughter. Swiftly, the musicians joined the bonfire. I squeezed between a singer and a flute. The music was loud and overpowering. The sound took control of the healing house. Conflict was diffused.

Later on, as the last round of yagé is served, I joined the line. The foreign guests were shivering in their hammocks in a frozen stupor. They all looked asleep, but I doubt it. I was told that "my new friend" is the group's guide. He organizes traveling experiences in "unconventional" places. "Ayahuasca in the Amazon" is probably one of his highlight tours. At dawn, as usual, almost every indigenous person in the house is talking loudly and jovially. Hammock to hammock conversations as well as loud general comments. We are alive. It's a beautiful day. We are surrounded by nature. And... no one got hurt. I was laughed at. *El que sabe sabe,* is the joke of the hour. I, too, was laughing, and healing. It's been a few years since that episode, and still today, every so often, I'm greeted with, *"El que sabe sabe,"* followed by a knowing look and a smirk. I am still wondering what the heck the joke meant and why people are still laughing at me.

What Now?

A couple of days after the events I described above, I asked a group of young yagé drinkers why I had not seen them at the ceremony. They answered that they didn't feel comfortable attending *ceremonies* in the presence of tourists. Their comment both worried me and gave me hope. The growth of spiritual tourism in the Colombian Amazon might interfere with resilient and essential local medicinal knowledge systems and practices and transform them in ways that we cannot predict. However, the fact that young yagé drinkers are concerned about protecting their practices gives me hope about the future. Three years ago, as the UMIYAC was revising its new strategic plan, a group of young Inga people made a case for implementing a youth program. This has now been running for two consecutive years, and it has grown to include communities from several ethnicities.

Indigenous spokespeople from all corners of Colombia never tire of saying that spirituality is the main pillar for the survival of indigenous peoples. They insist that only thanks to ancestral spiritual knowledge, their communities were able to endure colonialism, religions, and sequels of uninterrupted extractive invasions. Today, the market is aiming straight at this spiritual knowledge. A recurrent pitch is that ayahuasca, along with other psychedelics, must go viral to heal our sick world. Only then will people understand the value of indigenous practices and the importance of Mother Nature. When this happens, armies of spiritually activated humans will finally join nature's cause. Finances will pour into the communities, and we will all live happily ever after.

The healers, whilst somewhat participating in this burgeoning reality, are wary that indiscriminate commerce, medicalization, and neo-shamanism are stripping yagé of its curative spiritual powers. In commercial rituals performed by inexperienced practitioners, spirits, malignant or otherwise, roam around uncontrolled, inflicting long-lasting pain and causing more illnesses and eventually spiritual death. Bad practices by self-appointed healers and neo-shamans are seen as the cause of most of the incidents during ceremonies that have appeared in the news in recent years which include stabbings, deaths, suicides, people leaving the ceremony ahead of time and suffocating on their own vomit, sexual harassment, rape, and emotional manipulation.[18]

Whilst helping some families generate badly needed cash flow, the indiscriminate marketing of indigenous knowledge systems, rituals, and sacred plants, to an extent for example like that of Iquitos, Peru, might exacerbate community divisions, interfere with processes of local resistance, slow down collective healing and cultural revitalization, and hinder peace building and reconciliation.

Perhaps, people and organizations that want to advocate for indigenous peoples and be allies in their resistance should downscale all activities and research related to yagé/ayahuasca. Simultaneously, more energy and resources should go towards decriminalizing coca leaves and regulating cocaine markets in the global north with the aim of ending the war on drugs once and for all.

UMIYAC's relationship with international NGOs is an exemplary case study from which anyone can learn. In February 2023, UMIYAC and ASOMI, the organization representing women botanical healers, got together in a ceremonial space and performed an act of friendship and reconciliation. This was only the last of a series of attempts and steps taken to heal a profound division provoked and sustained by interfering outsiders.

International NGOs have a duty not to provoke divisions and to abide to strict Do No Harm protocols. This is especially important when interacting with communities living in war zones and at risk of extermination. Communities that, let's not forget, rely on unity as a primary survival strategy. Moreover, rural communities and social movements are getting increasingly restless with aid and development programs that are designed and decided upon in urban capitals and later presented as the only "take or leave it" alternative. The same thing is true for extractive research practices and non-transparent hyped viral "Save the Amazon" fundraising campaigns negotiated on their behalf.

Indigenous knowledge systems, spiritual knowledge and practices, together with resilient community adaptations, can help humanity solve problems related to climate change, war, ecocide as well as the widening mental and spiritual health crises. Moreover, Colombia and the Latin American continent possess a rich tradition of social methodologies and participatory research designed to foster autonomy, localized sustainable development, and self-determination. Communities and social movements are continuously perfecting and applying these methodologies. It is time for allies, academia, governments, and international NGOs to start listening very carefully and incorporate these inputs into mainstream theories and practices. This is the only way to unlock stagnant, hijacked institutions and transform them into viable agents of positive change.

References

1. Conversations with spiritual leaders, Colombia, 2010–2023.

2. Fordham, Gupta, 2011; Gingerich, Vitale, 2017; Nakashima, 2010.

3. Official Registry of Victims, 2018.

4. Piedra Alegre is a fictitious name used to protect the identity and location.

5. For more information about the legislation related to *resguardos*, see: Reguardo's Law: http://www.acnur.org/t3/fileadmin/Documentos/BDL/2008/6654.pdf?view=1

    Law 21 of 1991, which incorporates ILO C. 169 of 1989 into Colombian legislation: http://www2.igac.gov.co/igac_web/normograma_files/Ley21-1991.pdf

    Norms of Indigenous Resguardos DANE: http://sige.dane.gov.co:81/gruposEtnicos/doc/NormatividadResguardosIndigenas.pdf

6. Gratteri, N., Nicaso, A. (2022). *Oro Bianco*. Milan: Mondadori.

7. https://oxfamilibrary.openrepository.com/bitstream/handle/10546/294327/bp-colombia-contested-spaces-010313-en.pdf?sequence=1

8. Writer Johann Hari in *Chasing the Scream* argues a similar point.

9. *Raspachines*: colloquial term for coca harvesters.
10. *Tigreros*: *taitas* who through metamorphosis could transform into tigers and other animals.
11. Bia Labate.
12. Alhena Caicedo-Fernandez (2015). *La alteridad radical que cura. Neochamanismos yajeceros en Colombia.* Bogotá: Universidad de los Andes. https://www.jstor.org/stable/10.7440/j.ctt18pkdj8
13. Caicedo, 2009; Langdon, 2015; and Uribe, 2008, amongst others.
14. *El conocimiento*: the knowledge that a person acquires practicing yagé medicine.
15. Personal communication, Putumayo, Colombia, 2017.
16. *Chicha* is a fermented (or non-fermented) beverage ubiquitous in indigenous communities. It is generally made from sugarcane, corn, plantain or pineapple.
17. In 2017 the United Nations High Commissioner for Human Rights reported 441 attacks against human rights defenders and social leaders, including 121 assassinations. This is a 45% increase since the previous year.
18. *Guardian,* 18/12/2015; *NZherald.co.nz,* 22/10/2015; *Latin America Current Events & News,* 27/11/2011; CNN, 27/10/2014; *Colombia Reports,* 5/6/2015.

Bibliography

CNN (27/10/2014). Teen's Quest for Amazon Medicine Ends in Tragedy: https://edition.cnn.com/2014/10/24/justice/ayahuasca-death-kyle-nolan-mother

Colombia Reports (5/6/2015). Central Colombian Shaman Arrested for Drugging and Raping at least 50 Women: https://colombiareports.com/central-colombia-shaman-arrested-for-drugging-and-raping-at-least-50-women/

Fordham, M. (2011). Gupta, Leading Resilient Development: Grassroots Women's Priorities, Practices and Innovations. *GROOTS* and *UNDP*.

Guardian (18/12/2015). Briton Stabbed to Death by Canadian During "Bad Trip" on Hallucinogenic Ayahuasca Plant. https://www.theguardian.com/world/2015/dec/18/canadian-man-kills-briton-ayahuasca-ceremony-peruvian-amazon

Labate, Cavnar, Gearin (2017). *The World Ayahuasca Diaspora: Reinventions and Controversies.* London: Routledge.

Langdon, E. (2016). The Revitalization of Yagé Shamanism amongst the Siona: Strategies of Survival in Historical Context. In *Anthropology of Consciousness*, Vol. 27, Issue 2, pp. 180–203, ISSN 1053–4202, © by the American Anthropological Association.

Latin America Current Events & News (27/11/2011). Peru, French Tourist Dies of Suspected Overdose of Ayahuasca: http://latinamericacurrentevents.com/peru-french-tourist-dies-from-suspected-overdose-of-ayahuasca/13972/

Nakashima, Douglas (ed.) (2010). Indigenous Knowledge in Global Policies and Practice for Education, Science and Culture. Paris: UNESCO.

NZHerald (22/10/2015). Kiwi Traveller in Peru Dies After Amazon Drug Ritual: http://www.nzherald.co.nz/nz/news/article.cfm?c_id=1&objectid=11516673

Ruling 004 of 2009, Colombian Constitutional Court: http://www.corteconstitucional.gov.co/relatoria/autos/2009/a004-09.htm

United Nations High Commissioner for Human Rights (March 2, 2018). Annual Report of the United Nations High Commissioner for Human Rights on the situation of human rights in Colombia: http://www.hchr.org.co/documentoseinformes/informes/altocomisionado/A-HRC-37-3-Add_3_EN.pdf

14

The Case for Just Plain Tripping

Chris Kilham, the Medicine Hunter

Chris Kilham is a medicine hunter, author, educator and yogi. The founder of Medicine Hunter Inc., *Chris has conducted medicinal plant research and sustainable botanical sourcing in over 45 countries. Chris works with companies to develop and popularize traditional plant-based food and medicinal products into market successes. These include ashwagandha, kava, maca, rhodiola, schisandra, tamanu oil, cat's claw, dragon's blood,* ayahuasca, *and hundreds of other plants. Chris also works to bridge worlds, regularly sharing information about other cultures through presentations and media. The New York Times calls Chris "part David Attenborough, part Indiana Jones."*

Chris has appeared on over 1500 radio programs and more than 500 TV programs worldwide, with features in The New York Times, ABC News Nightline, Forbes, The Wall Street Journal, The Dr. Oz Show, NBC Nightly News, Good Morning America, ABC 20/20, Psychology Today, Outside Magazine, Newsweek, *CNN,* FOX News Health, LA Yoga, LA Weekly, New York *magazine,* Boston *magazine,* UTNE Reader, Playboy, VICE, Prevention *magazine,* Business Insider, Oprah & Friends, *PBS* Healing Quest, Woman's World, *and many others. He has additionally appeared on PBS,* CNBC Power Lunch, *CNN, MSNBC, BBC, NPR, CCTV, France 2, Chelsea Handler's* Chelsea Does, *and* The Preachers. *As a TV correspondent and guest, he speaks about medicine hunting, traditional botanical medicines, nutraceuticals, psychoactive plants, environmental and cultural preservation and other related topics for a broad and diverse variety of audiences. His latest book,* **The Lotus and The Bud: Cannabis, Consciousness, and Yoga Practice***, offers an in-depth guide to blending the practice of yoga with cannabis.*

∞

As the world changes so does the psychedelic scene, which is not actually a cohesive entity in any way, but more a vast dendritic network of diverse people utilizing psychedelics for a broad variety of purposes. In recent years we have endured endless promotional hype and pap from start-up psychedelic pharmaceutical entities promising to "usher in a new era of healing" and "radically transform the medical landscape" and "creating a whole new paradigm" blah blah blah. These and other fatuous claims must be taken with an immense grain of salt, perhaps even a full cow-licking block. Pharma companies have always made promises like these, whether they are hawking antibiotics or anti-inflammatories. It's part of the pitch.

Let's clear away some debris and get on point. For starters, most of the starry-eyed, big promise psychedelic start-ups are developing synthetic drugs, for which they hope to secure patents that will ensure a high rate of financial return for investors, and proprietary protection in the pharmaceutical market. It's the magic bullet approach to drug development, the *modus operandi* that pharma has always been about— big investment, development of novel molecules that can be patented, and expensive clinical trials that take years to perform, at the end of which FDA approval of their drug either does or does not happen. In the course of this slow-rolling folly, a couple of companies have even attempted to patent (without success) sitting on comfy furniture with eye shades and holding a therapist's hand. It's nut country out there. And in case anybody has not noticed, millions of people are tripping anyway, and few people but investors and those studying to be clinical trip-sitters are waiting for the drugs to roll out. Why is this? We already have a rich world of psychedelics.

I have predicted in a variety of articles and talks that many of the psychedelic pharma entities will die bleeding in the streets, and the fever for giant investment will cool in the face of the sober realization that this is actually the same old expensive and lumbering drug path, the *Hunger Games* with visionary agents. At the end of a successful development run, a psychedelic pharma company will need to charge obscene dollars relative to existing and well-established prices for psychedelics, and that high price will necessitate doctor visits, prescriptions, third-party reimbursement schemes, co-pay plans, all the garbage bureaucracy of the prescription drug scene today. For those who choose medical supervision and clinical settings to deal with traumas and troubles these grossly overpriced drugs administered in clinical places will likely help many.

The value of this cannot be overstated. But outside of the clinical silo, right now today there are millions of people consuming psychedelics that do not come from pharmaceutical companies, that have nothing to do with a medical scheme, that do not cost a fortune, and that enable people to exercise their basic human right of cognitive freedom. To the best of our knowledge, this kind of non-medical tripping has been going on for thousands of years.

We mostly know about psychedelics at all thanks to ancient and indigenous use of plants and fungi to alter the mind, effect thorough healing and crash the gates of heaven. The potent Amazonian botanical brew of ayahuasca, peyote, San Pedro cactus, various DMT-containing snuffs, morning glory seeds, iboga, psilocybin mushrooms and oral doses of cannabis have been utilized for centuries or millennia as full-blown psychedelics by indigenous people, especially those living in the so-called psychoactive breadbasket ranging from Mexico to the southernmost tip of South America, although iboga notably derives from Africa, and cannabis appears to have emerged from the last ice age in central northern Asia.

These agents all arise from nature, imbued with the intelligence of millennia of clever evolution, phytochemical complexity and genetic genius. Human beings have figured out how to identify, prepare and use these for journeys into the spirit world. That said the accidental discovery of the mind-blowing powers of LSD was in fact a laboratory fluke that came without any form of shamanic warning. Still, LSD was developed from lysergic acid, which in turn is derived from the highly toxic fungus ergot, or *Claviceps purpurea*, which enjoys a long history as a deadly food poison, a traditional aid to easing childbirth and a powerful agent utilized in the ancient mysteries at Eleusis in early Greece to induce divine visions. Acid may have been whipped up in the Sandoz laboratories, but at its core it is powered by the nuclear compound of a fungus which dates back at least 100 million years.

As we bear down on psychedelics a bit of definition is in order. There is but one defining factor of psychedelics, and only one. A psychedelic is an agent or preparation that can promote a mystical experience. Many substances, including alcohol, opium and Brugmansia, can cause visions. You can hallucinate like mad on various toxins and pathogens. But those do not lead a person to a mystical experience. A great many more agents, from cleaning fluids to barbiturates, will space you out. But they will not promote a mystical experience. With true psychedelics like ayahuasca, peyote, psilocybin mushrooms, San Pedro, LSD, DMT-rich snuffs, iboga and large dose oral cannabis, the one essential bedrock factor is that they can and often do generate a mystical experience. This frequently revolutionizes people's lives, generating a cascade of more life-imbuing understanding, attitudes and behaviors.

Mystical experiences have been very well described in both religious and nonreligious settings throughout human history, and involve common factors. In the mystical experience, the apprehension of one's personal self dissolves into cosmic consciousness, a wholly consuming universality, dissolution into an ocean of pure existence unfettered by identity, description or ideation. The self as we typically know it is wiped away, replaced by boundless energy and joy, often accompanied by a vision of brilliant, dazzling light, transcendence of time, overwhelming love and a profound compassion for all beings. Existence itself is experienced as limitless energy, suffused with bliss. To some, this idea sounds frightening. Losing oneself entirely to an experience of pure existence can seem a scary proposition. But for many who have enjoyed such a mystical experience with the aid of psychedelics, the event has been one of immense positive impact on their lives.

Mystical experience is not religion and this confuses some people. A religion by definition has a doctrine, while no doctrine is required to conjure a mystical experience. You do not need to believe any particular thing or subscribe to any notions. Ideas are not the experience. Words are not the thing. The map is not the territory. Religions invariably involve petitioning deities of various sorts, while mystical experience requires no such entry fee. Religions involve isms, ologies and osophies, always associated with a charismatic leader, while mystical experience requires none of those trappings. One is delivered to a sense of pure, ecstatic spiritual being, beyond notions of any kind, cosmic consciousness. The totality of our energetic being is our spirit, the unique animating force of all life. Many people name this spirit god or assign the identity of a deity to this. But that is an interpretive choice and not the essence of the thing. It is only after we have been blown to pieces by mystical experience that we set about to define and fashion that experience into words and ideas that may seem to make sense of it.

The plot thickens a bit when intense mystical experiences involve the invocation or worship of deities or are delivered in the context of a psychedelic religion like Santo Daime or the Native American Church. Some people encounter Jesus or Mary or Kali or Aphrodite, Buddha, Our Lady of Guadalupe, Maitreya and various prophets and spiritual figures. These experiences take place in contextual settings such as Christian or Buddhist venues, archaic goddess ceremonies, etc. In some cases, iconic figures spontaneously show up.

Unless someone is hewing to a specific religion or doctrine, who can say how a jolly Ganesh winds up showering your crown chakra with musical gold coins, or how the Green Man or the Elfin Queen show up? The workings of the mind are complex and endless. Maybe you find yourself sitting in full lotus inside the third eye of the Buddha, or are pulled by a tractor beam of love into the heart of the divine mother. Maybe you ride a cobra snake up the vertebral channel of the god Siva and then blow out his top knot and are dispersed into infinity. Anything can happen. Trillions of events precipitate these moments. But whatever the visions or phenomena that bring you to the point when you are immersed in the mystical experience, you wind up beyond all that phenomena, formless and boundless, dwelling in the essence of all being.

When my friends and I started taking LSD in the late 1960s it was still legal. My first three acid trips were in fact completely legal, though soon enough state and federal laws changed all of that. By mid-1968 that legal ride was over, though we soldiered on valiantly for years, dropping acid often and having a grand old time. For many of us acid was a gateway to meditation, yoga, environmental activism, nature exploration, global travel, cultural exchange and a deep dive into holistic health. We tripped unaided by kind guides who might otherwise hold our hands, without eye shades or purpose-designed furniture or lilting background music. There were no questionnaires or follow-up sessions. Nobody sat and took notes. We danced in the rain, laughed on beaches, sat around fires or swimming pools, ran about in forests, and had snowball fights at 2:00 a.m. while laughing in the winter cold and the snow sparkled like stars, with loud rock playing in the background. We sang the popular hits of various groups from the Beach Boys to Otis Redding as the sun rose, and occasionally flung ourselves naked into the ocean or a lake. We went to concerts and busted our conks on rock music that penetrated every nerve. And at some point or another pretty much every one of us had transformative mystical experiences that rocked our mortal souls and shifted the trajectory of our life paths.

For years I don't think it ever occurred to any one of us to consume less than a tripping dose. When presented with LSD or peyote buttons or mushrooms the question was always how much was required to trip. And despite our lack of medical supervision, virtually everybody I knew wound up better off for immersion into the mystic deep of psychedelics. Today we have psychedelic microdosing (just a bit), medidosing (a bit more) and tripping dosing, with many people going for the former tamed-down versions, which undeniably impart benefits. But if we pay attention to how we got here, we must tip our hats to native people who developed powerful tripping preparations and ceremonies, to Ken Kesey and the Merry Pranksters, whose full dose acid tests turned on thousands in the 1960s, to the Grateful Dead whose traveling caravan of acolytes tripped balls on acid, mushrooms and all manner of visionary agents all over this great land for decades, and the many psychedelic luminaries including Timothy Leary, Aldous Huxley and others who carried and kept the torch of full-on tripping burning bright.

Even as media rushes breathless into the rib-crushing embrace of the pharmaceutical psychedelic sector, millions of people have tripped and are tripping, swallowing or snorting or otherwise ingesting sufficient enough quantities of natural psychedelics to embark on a real journey of the soul. Right now 5-MeO-DMT-rich bufo toad exudate is all the rage, an atom-smashing experience residing far away from the custom lounge chair of a clinic. Removed from the world of drug development, smart, nature-derived psychedelics deliver sheer brilliance in their phytochemical complexity and genetic genius, promoting life changing revelations and immersion into pure spirit. Many is the night I have sat in a maloka or a hut someplace in the Amazon, dissolved into pure energy and bursting with love for all beings after drinking a large amount of ayahuasca, soaring along with spirit wind in my sails, lost in the croaking of thousands of forest frogs and moved by songs sung by shamans, ripping high and ecstatic. I never got a prescription, filled out a form for Blue Cross reimbursement or scheduled follow-up therapy sessions at any clinic at all.

In psychedelic plants and fungi we have extraordinary gifts from nature which have evolved over millions of years. We have co-evolved with them, sharing genes and physiology. They come to us each one brimming with hundreds of synergistic compounds, imparting mind-altering chemistry and genetic genius and a vast body of use dating back to antiquity. With the psychedelic drugs in development, that complexity, genetic intelligence and millennia of use are absent. The summation of this is a personal exhortation to seek refuge in the splendid complexity and genetic genius of natural psychedelics, including LSD due to its ergot-derived core. People don't just trip because they need to mend a broken mind. Most often they do so to enjoy a liberating, enlivening and often visionary journey into the far reaches of the mind, to the spirit essence of all things known and unknown. Mind the rules of set and setting, know your sources, get your dose right, take a swan dive into the deep end of the pool and trip. That's where the goodies can be found. As I like to say, go big or go home.

Section 3

Personal Transformation

15

Opening to the Visionary State

Simon Haiduk, Visual Artist

Growing up in British Columbia, Canada, Simon developed a strong affinity to nature with creative foundations in visual art and music. In 2004 he gravitated towards painting as a full-time endeavor bringing his musical background into the visual realm.

Simon has explored many visual mediums, often with a strong influence in spiritual themes connected to nature. In 2007 he graduated from Vancouver Film School, where he learned motion-graphics, among many other creative topics within the Digital Design program. That led him to animate many of his paintings now used in live VJ performances with various global music acts, including his own.

He continues to produce music by himself and collaboratively.

Within art galleries, festivals, conventions, and online platforms, his work is exhibited globally. He currently works out of his studio in BC, Canada, while continuing to explore an ever-expanding palette of creative endeavors. The cover of this book is his piece **Night Vision**.

∞

The first thing that comes to mind in writing a chapter for this book is a strong resonance with the title, *Infinite Perception*. Entheogens have for me been an amplifier for any previous inspiration while also blasting open the subconscious mind into realms beyond my wildest imagination. They have also brought focus, insight, connection, love, compassion, peace, empathy, wisdom, healing, and much more, as inspiration across the spectrum of my whole life.

Since I'm writing here about entheogens and their connection to my creativity, I feel that it would be fair to begin by saying that I've been very creative since childhood. My parents have always been a foundation of strong support with my creative endeavors. I was always into drawing with various mediums, then into music, picking up, learning various instruments, and other creative mediums along the way. My connection to lucid dreams, meditation and altered states of consciousness in various forms having a strong influence in the paths I've chosen.

I was exposed to entheogens, such as psilocybin mushrooms, LSD and cannabis as a young teen, mostly in recreational party settings. My early experiences often shed light on and exposed the awkward shadow sides of human interactions, while also giving hints at metaphysical energies. I had plenty of teenage angst, listening to punk rock music, with a desire to rebel against the "system." Many things I observed with how society operated seemed very confusing and I was drawn into realms of exploration to find better understanding. I knew there were alternative truths to the mountain of lies that we are given and meant to follow.

For me, visual art has been my primary method of attempting to describe mystical experiences. However, music is also one of my strongest allies in the creative process. Over my teen years and early twenties, I was much more into playing music than making visual art. I was in various bands and recorded my own solo albums, in a basement home studio. I didn't really like playing cover songs and was often expanding on the theme of experimental expressions.

Visual art was really catalyzed around this time by a series of five intentional LSD sessions over the period of five months, in 2003/4, when I was 23. In the first session, a friend and I were listening to psychedelic rock band Tool and it was then that I learned I could draw sound. An overwhelming sense of synesthesia emerged. My arm moved in an automatic way to all pulsations, beats and melodies of the music. Worlds upon worlds appeared, like a multidimensional M.C. Escher image made of woven sound wave scribbles. What I was seeing, seemed impossible, and to anyone viewing the image after, it could appear as some abstract mess. However, for me it was an absolute revelation in my perception of visual art connected to sound. My friend joined in on the drawing experience, as infinite realms of worlds upon worlds, weaving geometry, danced beneath our pens.

In the second session, there were four of us, again listening to Tool, with a general intention to draw on our journey. When the LSD kicked in, I picked up my brush pen and gently brought the tip to the surface of the paper. Immediately, my vision expanded into an impossible view zooming in on the microscopic drop of ink coming from the tip of the pen while simultaneously reflecting our room on the surface of the ink drop. There was an acute awareness that all of us were looking at the same thing, and to highlight the point we then all looked up at each other and spoke a collective "whoah," with eyes bulging wide in bewilderment for seeing the impossible, together. Continuing, slowly, each drop coming out of the pen revealed new words upon worlds within their surfaces while still reflecting our room. This kind of bewildering astonishment for what's considered normally impossible in senses of perception is a common aspect of psychedelic journeys. It's especially potent when it's simultaneous with one or more other person. Often understood with only a glance of the eyes, as psychic energy becomes more lucid.

It was after this experience that I took it upon myself to dust off the old paintbrushes and have my own solo intentional trip, as an experiment. I painted black over a previous unresolved painting, never before starting from a black background, it's just what I had available to work with. Beside me, as a kind of reference guide, was a magazine of cosmic images from deep space telescopes. This led to my first visionary art painting called *Astral Projection* wherein a figure sits near a tree, emanating electric glowing energy forming a long fractal portal into the distant cosmos. The space images in the magazine, along with the black foundation, inspired a new way of looking at things for me that brought forth a very electric almost neon kind of lighting, where the light came from inside the pseudo-forms. This was new and revelatory to me. I also hadn't been aware of much psychedelic art before that, besides computer fractals, which added to my feeling of revelation for something new.

It was soon after that my friend Castor, who exposed me to the music of Tool, and was part of these intentional LSD sessions, showed me the psychedelic cover art for one of their albums, *Lateralus*. It was created by artist Alex Grey. We went to Alex's website and on the front page was an invitation to go with him to Brazil for ten days of ceremony with the psychoactive plant medicine ayahuasca. The main intention being to *create art which evokes higher states of consciousness*. There was a photo of Alex holding a painting of an interdimensional being he met while on an ayahuasca journey. It was also bright neon-like colors on a black foundation, very similar in style to the recent new painting I had done. My being ignited with resonant passion seeing that there was someone else creating art like I had just experienced.

I wasn't able to make it on that trip to Brazil, however, at this time I had at least an understanding that people could supposedly astral travel while experiencing ayahuasca, so I had an intuition of what to do next. I arranged for another intentional LSD session with my friend Castor, to be during the same week that Alex and people were in Brazil with ayahuasca. My intention was to make some kind of astral connection, regardless of time and place. Though I had not told Castor about this detail, and it was more at this point in the back of my mind as a knowing that... something could happen.

For this journey, I had put my first painting *Astral Projection* on the wall of the small room we were in, and while reaching melting point, with the lyrics of Tool coming through saying "push the envelope, watch it bend," with each breath the portal on my painting receded into space right before our eyes. A jewel-like tapestry pattern moved upon every surface in the room as it pulsated in unison. We looked at each other in acknowledgement of this shared moment. Astonished. I was then forced by unseen energy to sit down, close my eyes, and it was then I felt an intense kind of heat, pressure, and tingling on the spot between my eyes on my temple. This area is also often referred to as the "Third Eye." Synchronistically, this references another Tool song called *Third Eye*, wherein the lyrics state, "prying open my third eye!" An apt phrase for this experience.

I rapidly transfigured into electric ether, out of my body, traveling through galactic portals, other dimensions, seeing past, future and present lifetimes; a seemingly infinite version of everything. My eyes cracked open and I encountered conscious orbs of light dancing on the now visible sound waves moving towards me from the direction of the speakers. Another realm of light stretched off into the seemingly infinite distance and the semi-materialized light orbs of consciousness moved as fiery morphing jewel eyes of pearlescent liquidity on an abstract grid of electric geometries. Another room now appeared juxtaposed with the room we were physically sitting in. People sitting in a ceremony circle, bodies of light, streams of blissful interconnected energy. I had suddenly remembered my intention for this journey, and it felt like I had reached the Ayahuasca ceremony in Brazil through my astral body. Upon returning to my physical body, after another journey through the cosmos, a benevolent light-being appeared showing me another version of myself as "asleep" in another dimension where I was being taken care of while I was having this human experience. It was a kind of reassurance. That other realm was so familiar, and in a way seemed more true than this physical world on Earth. A feeling that I had been here on Earth many times before became unwaveringly apparent. (Of course whether or not this is "true," it's a strong experience that leads to an unfolding of new awareness to be integrated in this life. That goes for any psychedelic journey.)

Recounting what had just happened, we were both not surprised that we had gone on this journey together. It confirmed this experience as something just as real as anything else in daily life. Especially since I had not told my friend my intentions of connecting with people across the globe through my astral body, yet it happened, and he came along for the ride. I sketched that evening and it soon after became my second visionary art painting ever, titled *My Awakening*.

That experience became a cornerstone for many deep journeys to follow, up to this day. I feel that recounting it, in some detail, gives insight into the transformative nature of psychedelic journeys, because after something like that, it's hard to go back to any kind of "normal." There's no turning back, and my art and music would never be what it is today without the influence of entheogens. I feel this is for the better.

It was the same year that many synchronicities fell into place like a falling trail of dominoes. Months later I met a group of visionary artists, called Tribe 13, at a psychedelic music festival.

I showed them some prints of my first recent paintings and they understood completely. Taking me into their tribe, exhibiting at festivals and galleries along the West Coast exposed to me the emerging culture for which visionary art plays a massive role. They were the first of many people I've met up to this day, who can see visionary art, based on psychedelic experiences, and understand it as a kind of language to describe the experience. It brings forth an awareness of something very familiar, understood beyond the constructs of physical being. An awareness of metaphysical energies which seem to be connected to the core construct of what makes physical reality. The underlying nature of our universe as light, sound and geometry. Modern quantum physics is also revealing aspects of this, theorizing many variations, with much research going into the idea of our universe as some kind of "holographic reality" —(for lack of better terms). Though, since words often fall short, this is where the art comes in. It gives people a chance to reconnect through eyewitness details of unifying realms that seem like God. I say God because there is often a sense of that kind of majesty, like aligning and bearing witness to the original creator. The beauty and the horror. It's easy to see the propensity for people to express the more heart-centered realms of light and love, over those of disgust and purgatory. Every psychedelic journey, for me, shows both the light and dark, revealing the contrast of all existence. This can seem like both a blessing and a burden within the continual paradox of life. There comes an understanding of choice as to which spectrum to focus on with integrated awareness.

Something else I've found fascinating which ties in here, is that when I've read reports of Near Death Experiences, ones where people go to other realms, there are often similarities to psychedelic journeys in visuals and feeling. To me this is more evidence of how entheogens access realms of original creation, or God. I and many people report having experienced death on these journeys. One example for me, I was in absolute horror while my physical aspects decayed, only to then be immediately held by golden webs of light surrounded by spirits looking down upon me like a baby in a cradle. Threads of warm true love, connecting us and everything. Again, with such a familiar warm feeling; being beyond physical space and time.

If science looks to find evidence of repeatable experiments, I would say that visionary art provides massive evidence for similarities in what people experience in these states. The similarities within the abstractions and specific details is astounding, and no coincidence. The amount of people who tell me that my art shows something almost exactly like their experience is staggering. Transformational festivals and the Internet have allowed artists to better connect and share notes as artistic renderings of these eyewitness accounts for further confirmation.

While that is amazing and something not to ignore, it was within the first two years of painting cosmic visionary art that I realized it was mostly only connecting with people that had psychedelic experiences. I felt a desire to bridge my new understandings with something that was more grounded and could relate to a broader audience, even if they hadn't done psychedelics. I wanted to use it as a tool to communicate and *evoke higher states of consciousness*. Having an understanding of the feeling and emotional effect art can have, I realized that nature is something most people can relate to, and certainly has been a primary source of inspiration for my whole life. This is where I got back into painting nature, with a new lens.

It became obvious very soon into painting with nature as a subject, still within a strong metaphysical theme, that more people could relate to it than my more cosmic images. In the early stages trees were the main subject. I love the detail of tree trunks, while they also provide physical and symbolic aspects of interconnectedness from the forest floor to the ether and energy, in the sky. Almost like an antenna to the cosmos. If there's a plant humans connect with more than others, I'd say it would be trees. After some years painting mostly forests, I started to incorporate wildlife. Animals are obviously an easy bridge for us to the nature realms. We have them as pets, and often imagine ourselves as one. I give much credit here to my partner Jane who has inspired much insight into the ways of animals. Her knowledge of them as spirit allies and their meaning in native wisdom traditions, as well as connecting to the *feeling* of the look, facial and body expressions, has been expansive and key to my evolution with this art. With emphasis on the feelings that allow people to see the image as a kind of mirror for something we know within our soul which connects to nature. Perhaps it's a more simple and symbiotic life with no ownership and attachment to things.

Of course I couldn't leave out my relationship to metaphysical energies with light and sound as the source energy. Along the way I've learned many skills with various computer software which has allowed me to incorporate these aspects by including animation and music. It's often been hard to decide on a still image anyway, so this multimedia approach allows for more layering, with motion and music, to give a better rendition of the metaphysical experience from my perspective. I have an ongoing series of artworks called Mettamorphs, which incorporate this approach. Some of these combine my own music, or are collaborations with artists of similar resonance to nature such as Yaima and Anilah. This is where I recommend you having a look to better understand.

My connection to the plant world, through plant medicine entheogens, is a blessed gift that I cherish. Being able to have an experience of becoming moss, or linking my consciousness to mycelium for access to the Internet of the forest, among many other insights, has provided deeper understandings of my place on Earth. There is added a mission of observation and integration to steward the land for more harmonious relationships with all beings here.

As I write this, 14 years after prying opening my third eye, having "my awakening" and becoming established in the global visionary art community, my use of entheogens is limited to maybe once or twice a year. Sometimes a microdose while making art, or a large dose for a deeper journey. It has been this way for many years now, as I continually turn towards cultivating grounded connections of kindness and compassion, with people and the Earth. I continuously move towards a more holistic approach to life, with art as a tool of expression, and my garden as a temple for Earth connection. Nourishing my body as a vehicle of the soul, temporarily passing through this world, one of many infinite realms, offering gifts of beauty from the never-ending source of inspiration that plant medicines provide.

16

Things That Huachuma Taught Me

Scott Lite, Ethnobotanist and Founder of EthnoCO

Scott Lite has studied plants and their complex relationship with humankind for almost two decades. He is an amateur ethnobotanist, anthropologist, herbalist, naturalist, seed-saver, plant hunter and apprentice to the shaman of the world. He studied herbalism at the Appalachia School of Holistic Herbalism.

Scott is an associate of the Botanical Preservation Corps and was on the 2010 expedition to Peru collecting seeds and information on medicinal, edible and sacred plants throughout the Andes. Scott has also participated in or organized and led other expeditions in the Amazon and Andes regions of Peru. Working directly with the indigenous people of Peru including the Quechua and Machiguenga his company EthnoCO sells fair trade goods and fields adventure expeditions in South America.

The Ethnobotanical Conservation Organization or "EthnoCO" for short offers Ethnobotany adventures, Cultural tours, Anthropology lectures, Herbal classes, Amazonian expeditions, Andean trekking, Bushcraft workshops, Volunteer projects, Outdoor education, Artisanal handicrafts, Traditional textiles, Rare plants/seeds, Natural products and info about biology, botany, shamanism, natural remedies, indigenous/native people and more.

The Sacred Cactus of the Andes — "San Pedro"/"Huachuma"

The four winds blew, the river whispered, the mountains called, and I had to go. When I was 20 years old, I made my first visit to the sacred land of Peru. I was on an expedition with the now defunct Botanical Preservation Corps to collect seeds and study indigenous culture. However, what really brought me, instructed me to go to Peru, was a magical cactus known as "San Pedro."

"San Pedro" is a sacred plant medicine of the Andes. "San Pedro" has been used for thousands of years in Peru, Ecuador and Bolivia as a shamanic medicine. The magical cactus of the Andes is known by many names: "Achuma," "Aguacolla," "Gigantón," "Huachuma," "Hahuacolla" and more. "San Pedro" means "Saint Peter" in Spanish, and it was given this name because Saint Peter, in Catholic mythology, is considered to hold the keys to heaven, just as the "San Pedro" cactus does. The old name, in the ancient tongue, is "Huachuma." The antiquity of this word is so great that it likely originated in a proto-Andean language, but it reached us through Aymara where its meaning relates to imbibing drink and altered states of consciousness.[1]

For clarity I will now use "Huachuma" to refer to the sacred Trichocereus cactus of the Andes.

What Is "Huachuma"?

"Huachuma" is the magical cactus of the Andes. Since before the beginning of recorded history, deep in our archaic past this sacred plant has been used in the Andes Mountains, its native habitat. Huachuma cacti grow throughout the Andes, often at formidable altitude. It is a tall columnar cactus growing up to 20+ feet high. Its skin ranges from light green to sky blue. The cactus consists of 4–15 ribs from which protrude spines, ranging from long and sharp to almost nonexistent. Mature plants produce colossal hairy green pedicels that reveal enormous white alien-looking flowers which are followed by bright green fruit filled with hard black seeds nestled in delicious cream-colored flesh.

The botany of Huachuma is slightly confused. All throughout the Andes various Trichocereus (syn. Echinopsis) species are utilized to make a range of medicines. Trichocereus bridgesii, Trichocereus pachanoi and Trichocereus peruvianus are the most highly regarded for use as an entheogen with specific strains or cultivars holding particular prestige.

All over Peru, today and in the past, Huachuma can be found growing in the homes and gardens of people from all social, economic and cultural backgrounds. Many modern Peruvians who grow the plant say that they keep it just for "good luck" or "protection." It is used "to protect homes," "as if it were a dog." They are unaware of its magical powers, great antiquity and immense role in the roots of Andean culture.

"Huachuma," Meeting My Ally...

It all started in the forest. As a young boy I was always fascinated by the mysteries of nature. I would go to the forest, alone, and watch the birds, the clouds, the river flowing. I would catch toads and crayfish. My father was trained in forestry. He was, and still is, quite the woodsman. He taught me the names of the trees and the birds. He showed me peace and wisdom sitting next to the babbling stream. I had found my temple, now I needed my sacrament.

This came some years later, during my first trip to South America when I was still quite young. It was there that I first met my friend and ally, the magical cactus known as "Huachuma." Sparked by my love of nature I began to read extensively about psychoactive and entheogenic plants. I was aware of the illegality of the plants in much of the world so I was excited when I found that, in South America, I could obtain the plants fairly easily. I searched the markets and eventually I obtained my first few "San Pedro" cacti, dull green and desiccated—right around my 15th birthday.

I didn't have a shaman, I didn't have a guide, I just had the scant information I could find in books and on the Internet, but I was determined, I was called to do this, and now, I had the cactus.

As there are no true, traditional Huachuma shaman left, as there were in ancient times, I had to teach myself. So I learned from the plant itself, visiting the ancient sites where it was used (Chavin de Huantar and others). I learned from the literature and direct practice.

My first experiences were not much to speak of; I didn't know how to prepare it, and I didn't know how to cook it. I got sick, threw up, felt a little weird but that was all. I tried again, another failure—*barf*. I froze it, blended it and tried all manner of preparation methods to turn this slimy green cactus into medicine—*PUKE*.

I tried again and again and again. I would perform chores and odd jobs in order to purchase more cacti to add to my growing garden. I was determined to find what magic this plant possessed. I learned, I read, I was quiet, and I listened to my heart and I listened to the voice of the cactus, which at the time were but faint whispers. Then it happened... after a long night of cooking, I awoke the next day and drank the medicine. I felt strange... was this it? The strange feeling became stronger and then fully took hold. I was ON "San Pedro" (as I referred to it at the time).

A sense of the sacred came over me, awe filled every fiber of my being as my eyes widened to see beams of sunlight stop in mid-air and shatter into a million iridescent crystals. Plants and fungi eagerly sprouted from the ground before me, and above, the trees were a kaleidoscope of infinite unfurling leaf, light and dappled shadow. I had a *vision*. I met my ally for the first time. I was stunned by its power and beauty.

It was a while before I took Huachuma again, but I wanted to know more. I wondered: who were the first humans to discover the magical cactus? What was its history and relationship to ancient cultures?

Earliest Origins: Guitarrero Cave

The origins of Trichocereus usage stretches back into the mists of Prehistory, so far back in fact, that our first evidence of Huachuma usage comes from one of the oldest sites in South America. The history of the Guitarrero Cave is so ancient that its earliest artifacts do not date to the Archaic but are classified as part of the Lithic period—likely even predating agriculture.

In central Peru, on the side of the Callejón de Huaylas valley, in the Yungay Province of the Ancash region, up the hill from the town of Mancos lies a mysterious cave: the Cave of the Guitar Maker, Guitarrero Cave. A 2015 EthnoCO expedition to the nearby town of Mancos found the cave above the Rio Santa at 2580 meters. One of EthnoCO's most experienced expedition team members Lorene R.'s notes from the field read as follows: "We saw almost nothing except cow's dung, spider webs and graffiti in the cave."

However, they did see some possible evidence of human presence, she continues in her notes: "We observed some red ochre on the walls of the cave but couldn't be sure if it was recent or thousands of years old." She also reported not remembering any cacti directly in the vicinity of the cave but cacti were spotted often in the region. The visit to the area didn't yield many answers but the view of Mount Huascaran (Peru's highest mountain) from the cave was splendid.

Known as "Cueva del Guitarrero" in Spanish, the enigmatic Quechua name is "Kitarawaqachiqpa mach'aynin" which means "get drunk while playing the guitar." It is one of the most intriguing sites in all of South America. The Cave of the Guitar Maker contains the oldest archaeological remains of crops that would sustain the Andes for thousands of years to come. This includes Peppers (Capsicum spp.), Beans (Phaseolus spp.), Eggfruit (Lucuma bifera), the potato-like tubers of Oca (Oxalis tuberosa) and Ulluku (Ullucus tuberosus), Squash (Cucurbita sp.) and possibly the very first variety of Maize/Corn (Zea mays).[1]

Some of the contents of the cave were found to be over 12,000 years old! Other items found in the cave include arrow/ spear points, rope, basketry, wood and leather tools. The cave also includes textiles — the oldest known in South America.[2] However, the most interesting thing found in the cave may be some of the ritual items; among them were remnants of Huachuma.[1]

Due to the presence of physical evidence of psychoactive plant material in relation to ceremonial items found in the cave perhaps the Huachuma at this site is the oldest, strongest evidence we have of entheogenic plant usage deep in early human history.

It is especially impressive that signs of the Trichocereus cactus existed in the cave for almost 10,000 years. We may have older evidence of entheogenic plant usage in Northern Africa at Tassili N'Ajjer, Algeria, but the evidence is circumstantial, in the form of cave paintings that are subject to interpretation.[3] The evidence from Algeria is inconclusive unlike the "Huachuma" found in the Guitarrero Cave; there is no physical evidence at Tassili N'Ajjer, thus the Huachuma found in the Guitarrero Cave is perhaps the strongest, oldest evidence for extremely ancient usage of entheogenic plants.

The Guitarrero Cave represents the most ancient origins of Andean culture and the progenitors of the great civilizations to come: Caral, Chavin, Moche and finally, some 8000 years later, the Incas. Some researchers have theorized that the people of the Guitarrero Cave are possibly the direct ancestors of the people who created the temple of Chavin de Huantar. The people of the Guitarrero Cave carried their ancient cactus-imbibing shamanic religion for millennia, resulting in a vast population of devout followers, finally culminating in the grand temple at Chavin de Huantar, a cactus vision carved into stone, manifested from the very Earth itself.

Chavin de Huantar: The Cactus Temple

The revered site of Chavin de Huantar in central Peru near the city of Huaraz was the epicenter of ancient Huachuma usage. Huachuma played a key role in the religion and rituals of the people of Chavin de Huantar.

The Temple of Chavin de Huantar must have truly been a wonder to behold with its grandiose festivals, mysterious rites, fantastical therianthrope Gods and court of charismatic wizards, priests and shaman. People came from all over what is now Peru, perhaps even further, to visit the revered temple. Much like the Rites of Eleusis in ancient Greece influenced Western Culture, the temple at Chavin de Huantar and its rituals played an integral role in the development of Andean society.

The "Temple of Doom," from the *Indiana Jones* movie, is based on the labyrinths of Chavin de Huantar. The earliest evidence of occupation of the Chavin site stretches back to 3000 BCE, with ceremonial activity occurring mostly around 300 BCE, the beginnings of the Temple itself date back to 900 BCE.[4]

The temple at Chavin is not nearly as ancient as the Guitarrero Cave but it makes up for it in grandeur. The Temple of Chavin de Huantar is made up of large rectangular buildings flanked by open plazas that display carved stone gateways and panels containing images of anthropomorphic eagles, snakes and jaguars. The immense main temple also had another curious feature: stone heads known as "Cabezas Clavas" depicted the process of psychoactive plant usage and shamanic transformation. Underground, the temple held vast maze-like labyrinths where initiates would be taken through rituals, ceremonies and ordeals by the shaman and priests of Chavin.

The initiate would be given a Spondylus shell of Huachuma brew (possibly with Brugmansia sp. or other plants added) to drink and may have also had "Willka"(Anadenanthera spp.) snuff blown up their nose by a priest. They would then be led to the underground labyrinths. Chavin was designed in such a way that channels of water flowed through the subterranean passageways creating a constant roaring sound that intensified as the initiate went deeper and deeper inside the underworld maze. The labyrinth was designed to have special acoustic properties.

Holes in various chambers would allow the priest to whisper incantations or blow a "Pututu" (shell horns) which would fill the space with reverberating sound. After stumbling through the darkness, now fully under the effects of the magical cactus, the initiate was worked into an ecstatic state by the plants, the rushing water, the disembodied whispers of the priests and the call of the "Pututu." Then the initiate would turn the corner, and out of the pitch blackness, bathed in blinding white light was the terrific, half-man, half-beast fanged God of Chavin carved in stone. It must have been a truly potent experience.[5]

The monolith mentioned above is 4.5 meters of carved stone. It's known in Spanish as the "Lanzon" and thought to be a representation of the main deity of Chavin. A fierce fanged anthropomorphic Caiman (relative of the alligator) is housed in the depths of the tunnels underneath the temple. Two carved stones clearly represent Huachuma. The "Stella Raimondi" depicts a shaman, holding two Huachuma cacti transforming into a fractal caiman, which can be viewed from either above or below, creating a sort of double image. "Estela del Portador del Cactus" ("The Cactus Carrier") or simply "The Shaman" is found carved into a stone panel in the temple's circular plaza. The man is clearly holding a Trichocereus cactus.

For hundreds of years the priests of Chavin de Huantar carried out their sacred rituals and ceremonies. It was a powerful cultural force and had major influence on the history, politics, religion and economics of the region—even vast distances from the temple itself. The priests and shaman of Chavin were highly influential, they held this power through religious reverence and inspiration—not military might. Yet one day, for reasons unknown, this power began to wane, the people moved away and the ceremonies slowly abated until they were all but forgotten... but some... some remember, and others are learning.

Huancabamba: The Village of Witches and Modern "Huachuma" Usage

Huancabamba lies in the misty eastern foothills of the Andes at around 2000 meters above sea level just south of Ecuador's border with Peru. Huancabamba is also known by its nickname: "Pueblo de los Brujos" or the "Village of Witches." It has held a special place in Peruvian culture for centuries, being a center of magic, spirituality and ethnobotanical plant usage. Huancabamba is known all over Peru and beyond for its powerful "Curanderos" (shamanic healers) and "Brujos" (witches).

People from near and far with every manner of disease and ailment come to the Village of Witches to seek a cure when modern medicine has failed them. They seek aid with health, work, love and luck. Others visit the mysterious town to have curses removed, find lost objects or cast a spell for the return of an estranged lover.[6]

Lagoons, situated above the town, are considered to be sacred. For hundreds, if not thousands of years, the lakes and lagoons above Huancabamba have held a special spiritual significance. Sitting at an altitude of almost 4000 meters the lakes and lagoons of "Las Huaringas" have many names. The best known lagoons are the "Laguna Shimbe" and the "Laguna Negro." In order to access the lakes one must be accompanied by the wizards who guard the lakes and dedicate themselves to magic and witchcraft—some light, some dark.

The healers and witches of Huancabamba use a plethora of plants, herbs, animal products and other substances for rituals and healing. The most important allies are "Tobacco" (Nicotiana tabacum/N. rustica), "Misha" (Datura/Brugmansia), Coca (Erythroxylum coca/Erythroxylum novogranatense) and of course the magical cactus Huachuma (Trichocereus pachanoi).

In the region of Huancabamba and beyond Huachuma is sometimes cooked with other plants to create the "Cimora" brew. Commonly combined with a large hanging flower known as Brugmansia, this powerful (and dangerous) brew is considered to be stronger than Huachuma alone. "Cimora," more common in ancient times than today, is still made in some remote areas of the Andes, especially northern Peru and southern Ecuador. Keep in mind that while Huachuma is safe, other additives in the "Cimora" brew can be dangerous, even deadly, if ingested in large amounts.

In the village below and the lagoons above ceremonies are held almost every night by one shaman or another. Often participants will drink the medicine in the evening, trip through the night, then walk to the lagoons at first light. They then strip naked and jump into the freezing, dark waters of "Las Huringas," the sacred mountain lagoons. Normally the participants in modern ceremonies are given mild, sometimes imperceptible doses of Huachuma. It is believed that simply ingesting the medicine heals you and is not necessary for a strongly psychoactive experience, though this is not always the case. The songs, rituals, bathing in the sacred lagoons and other ceremonial acts are considered most important.

Through the night a vigil is held in the home of the shaman and at dawn a trek is undertaken, climbing almost 2000 meters, up to the sacred lagoons. The shaman then asks those who have drunk the magical cactus their name, occupation and other personal details. Then the shaman begins to chant and sing vigorously over the patient in order to heal and remove bad energies. He may spit "Agua de Florida" (a scented alcohol concoction), wave swords or yell, sing, whistle and chant in order to drive away evil spirits.

As "Inti" (Sun in the Quechua native language) begins to warm "Pachamama" (Mother Earth) the sacred lagoons are finally reached. The participants may be whipped with stinging nettles (Urtica spp.) or have a "cuy" (guinea pig) rubbed over their bodies in various diagnostic rituals and healing ceremonies. They may be given tobacco juice which is snorted into the nose using a shell as a funnel before being commanded to jump into the freezing waters of the high mountain lagoon. The patient thus finds themselves healed from that which afflicted them. Be it through the power of the chemicals they ingested, evils spirits leaving their body or simply because they believe they are healed.

Huancabamba has its darker side. It's not only famous for those who cure but for those who curse as well. Not all the practitioners in Huancabamba work in the "light." For a small fee, with just a personal item, a photo and the name of the victim, the "brujo" (evil witch) can curse anyone, anywhere in the world. Often a piece of hair, a comb or toothbrush is stolen by the one who wishes to curse the victim and these items are brought to the "brujo." The "brujo" then performs his dark art and through sorcery wills some terrible event to befall the victim, be it illness, an accident, depression or the loss of finances.

The Witches of Huancabamba, the curanderos and shaman that live there are the last of a muddled lineage that started in the Guitarrero Cave over 10,000 years ago, blossomed in Chavin de Huantar more than three millennia ago and were carried through to the modern day. However, during the Conquest and even now the people who work with these Sacred Plants are often ill-treated. They endured Spanish oppression for hundreds of years in which time the religion based around Huachuma was changed, morphed and molded to be less offensive to the Catholic conquistadors.

In this process, the Huachuma ceremony, which was related to the Catholic Saint known as Peter, acquired and absorbed many aspects of the Catholic religion including its modern name "San Pedro."

The modern Huachuma ceremonies are deeply influenced by outside traditions such as Catholicism. The witches of Huancabamba often invoke the names of Jesus, as well as praying to the Saints and Mary, complete with statuettes and crucifixes displayed on the ancient Andean "Mesa" (ceremonial cloth).

However not all of the original native tradition is lost. They also call out to the ancient Gods, the "Apukuna" (Mountain Gods), "Pachamama" (Mother Earth), "Inti" (Sun), "Quilla" (Moon), "Yakumama" (Mother Water) as well as a plethora of other lesser deities. The modern Huachuma ceremonies held in Huancabamba do not represent the long-lost tradition of Chavin de Huantar or even of later cultures. It's a modern amalgamation of the ancient Huachuma ceremony blended with Catholicism and Spanish ideology. Only kernels of the ancient rituals remain… yet there are those who have attempted to resurrect the tradition.

The Ceremony

Walking the mountain trails. Cleansing in the sacred lagoons. Visiting the ancient temples. Offerings of coca. Harvesting in the hills. Cooking the medicine for days, nights, weeks. Fasting, eating rice. Drinking, wretched bitter slime. Puking. Learning. Retching. Forgotten profundity. Thanking Pachamama. Pricked by a spine. Insight. Time passes. Lessons learned. When I first began, I would always drink the medicine alone, but eventually, after many years, slowly at first, I would invite others to join in the ceremony.

The ceremony takes place in southern Peru at a special cave. This cave is where young Inca elite boys were initiated to become men. There they would take psychoactive plants, wrestle, learn new skills, race, listen to lectures from the old wise men, dance, recite poetry and take part in various rituals and ceremonies to show them how to be men.

Today, in the same sacred place, we start the ceremony with a passage called the "Plant Medicine Sutra" adapted from an old Buddhist text. We thank "Pachamama," the Sun, the Moon, the Stars, the Apus (Sacred Mountains) and Huachuma itself. Then we drink down the bitter medicine while remembering, "Good medicine tastes bad!" After a few hours of laying in the forest waiting for the medicine to take hold, we then walk a short distance to the entrance of the cave/temple. We must climb down some treacherous rocks (it's all part of the trial) as we enter a large labyrinth inside the cave with a river flowing through the center.

We sit, meditate, sing and think as we enter deeper into the sacred space. We feel roots grow from our feet into the Earth, light shining from inside our heart, through our head, radiating warmth in all directions. We vibrate and echo with the bones of the Earth.

One by one I ask the participants to follow me towards the exit of the cave where it darkens at a narrow passage. One by one they come forth, standing in the river they close their eyes. Whistling, chanting, blowing mapacho smoke, cleansing the energies, singing, cleaning, shaking the rattle, removing that which is not needed. Leave it in the cave. Leave whatever you don't need in the cave! Just as Mother Earth changes the manure of the cow into fertile soil "Pachamama" will recycle our bad energy, reuse it, make it new, changing it into something useful!

With a final last intense shaking of the rattle and blowing of mapacho smoke, "ZZZzzzSSSssshuuuuufff!" I exclaim. "You have been cleansed! Walk into the light!" The participant walks through the cave, slowly, alone, through the river with its freezing waters, inside the Belly of Pachamama, our Mother, through the darkness and into the light of a new dawn. Rebirth.

We then wait until the last rays of "Inti" (the sun) shine, leaving the forest in shadow and slowly head back to the sacred city, Cusco. We enjoy some light food while we talk of the wonder and the magic of the day, of the profound insights we obtained, of the friends we made, and of the things we learned. When we are ready, we depart, say our goodbyes and head for bed. Yet it would seem I cannot yet sleep?

Things Huachuma Has Taught Me...

Huachuma has taught me many things and given me many insights but perhaps the most valuable lesson I learned from the magical cactus is as follows: We (humans) must become the stewards of the forest. The protectors of the trees and rivers. The keepers of truth, light and love. Humans are the only living things on this Earth that have the ability to alter their environment in a profound way. We must alter our environment in a manner that makes it a better place for both plants and people, for humanity and the vast web of other life on this beautiful gift we call Earth.

Huachuma taught me to love myself but above all it taught me I am a warrior of light, one of the Gardeners of Eden and it is my Sacred Duty to protect our mother, Mother Earth, **Pachamama**. It showed me how humanity could turn this planet into a hellscape, devoid of life, scorched like a desert by mankind's arrogance OR how we can turn the Pacha (Earth) into a new Garden of Eden with mankind as her stewards, dutifully protecting that which grows, crawls, swims, walks or flies!

This is my Sacred Duty given to me directly from the very Earth itself by one of her most powerful messengers, **Huachuma**, one of the plants of the Gods! Won't you join us? Won't you heed the call?

I am not a shaman. I am not a curandero. I *am* a Huachumero, one who works with Huachuma. I am a Steward of the Forest, a Gardener of Eden. Reviver of traditions. I am blessed, I am thankful. Thank you, Huachuma, for your immense lessons, thank you for the guidance you have given me. Thank you for making me believe in myself again. Thank you for making me believe in the Universe again. Thank you, Pachamama, for feeding us, giving us the plants, giving us soil to grow our food and our medicine. Thank you, God. Thank you for life. These are the things Huachuma taught me. Thank you for reading my story.

I have not always known these principles, I had to be shown the way, guided by Huachuma. Huachuma showed me the path and illuminated it before me. It instructed me, "Help plants, help people, help plants"—this is my motto and this led to the foundation of my organization, EthnoCO (Ethnobotanical Conservation Organization).

EthnoCO is an organization based out of southern Peru. We work directly with the Machiguenga and Quechua indigenous peoples selling their wares and offering expeditions to their communities to work with the shaman, medicine men, healers, mystics, elders and keepers of ancient knowledge. Our goal is to connect cultures and offer people a deeper experience in Peru than simply taking a photo of a llama and visiting Machu Picchu. We aim to connect plants and people, and in the process, hopefully, make the world a better place.

A Note on Legality

Trichocereus live plants are legal throughout the world to own as ornamental or botanical specimens. I suggest everyone grow one for their beauty and luck-bringing properties. However, it is ILLEGAL to process the plant in any way (basically it is prohibited to do anything but just grow/propagate the living plant). I do NOT suggest anyone break the law. I DO, however, suggest visiting Peru if you are truly interested in working with Huachuma. All Huachuma usage and ceremonies that were discussed in the article took place in South America where this sacred medicine is completely legal for religious/spiritual usage. The author does not recommend anyone partake in Plant Medicine of any kind unless they can do so legally and safely.

References

1. El Cactus San Pedro: su función y significado en Chavin de Huántar y la tradición religiosa de los andes centrales. Leonardo Feldman Gracia. Dra. Ruth Shady Solis, p. 144. Link: http://cybertesis.unmsm.edu.pe/bitstream/handle/20.500.12672/2346/Feldman_dl%281%29.pdf?sequence=1&isAllowed=y

2. Chronology of Guitarrero Cave, Peru by Thomas F. Lynch, R. Gillespie, John A.J. Gowlett, R.E.M. Hedges. Link: https://www.researchgate.net/publication/6062471_Chronology_of_Guitarrero_Cave_Peru

3. http://en.psilosophy.info/the_oldest_representations_of_hallucinogenic_mushrooms_in_the_world.html

4. https://www.researchgate.net/publication/287773510_Context_construction_and_ritual_in_the_development_of_authority_at_Chavin_de_Huantar

5. *Chavín de Huantar. El Teatro Del Más Allá*. National Geographic Documentary, 2015.

6. Sacred Plants of the San Pedro Cult. E. Wade Davis, *Botanical Museum Leaflets*, Harvard University, Vol. 29, No. 4. https://www.jstor.org/stable/41762855

Other Sources used for this Article

Armijos, Chabaco et al. (2014). Traditional medicine applied by the Saraguro *yachakkuna*: a preliminary approach to the use of sacred and psychoactive plant species in the southern region of Ecuador. *Journal of Ethnobiology and Ethnomedicine*: https://www.ncbi.nlm.nih.gov/pmc/articles/PMC3975971/

Burger, Richard L. (1993). *Chavin and the Origins of Andean Civilization*.

Cracking Cryptocacti. *VICE* at: https://www.vice.com/en/article/avn73g/cracking-cryptocacti-0000202-v19n5

Davis, Wade (1996). *One River.*

Evans Schultes, Richard; Albert Hofmann; Christian Rätsch (1992). *Plants of the Gods: Their Sacred, Healing, and Hallucinogenic Powers.*

Lynch, Thomas (1980). *Guitarrero Cave in its Andean Context*: https://cuevasdelperu.org/publicaciones/peru/1980_guitarrerocave_lynch.pdf?fbclid=iwar3pyvxkiwqz59s0tytvzux-9aqol5nutqbdijqamv_ihfybkmxceyks7dry

New York Times (1985). Science Watch; Peru cave artifacts are 10,000 years old. This is a digitized version of an article from the *NY Times'* print archive, before the start of online publication in: https://www.nytimes.com/1985/09/24/science/science-watch-peru-cave-artifacts-are-10000-years-old.html

Sharon, Douglas, PhD (2015). *Wizard of the Four Winds.*

Voogelbreinder, Snu (2016). Garden of Eden.

Weber, George (2007). —Guitarrero cave (Ancash, Peru). Possible Relatives in the Americas.

17

The Will of Consciousness Revealed by Ayahuasca

Paula Rayo, Researcher at King's University College

Paula Rayo graduated from King's University College at the University of Western Ontario in 2016 with a BA honors in psychology and a minor in philosophy. Paula's main research interests are in consciousness studies, humanistic and positive psychology, and existentialism. Her honors thesis investigated the potential of ayahuasca shamanism to improve existential meaning. Paula is now completing her Masters of Professional Education in Curriculum and Pedagogy to implement new strategies for holistic learning in the education system.

∞

If you want to find the secrets of the universe,
think in terms of energy, frequency and vibration.
– Nikola Tesla

When we live life on autopilot, we let ourselves get tossed around by the ups and downs of life without any conscious awareness. We develop automatic and unconscious emotional reactions to the happenings of life, which often leads to a victimhood mentality. "Woe is me" is the commonly internalized mantra as we are constantly blaming everything and everyone else for the circumstances of our lives. That was the story of my life until I drank ayahuasca and I finally understood that if I want to make things better for myself, I must take full responsibility for the circumstances of my life. Ayahuasca taught me that we are constantly creating our circumstances and realities through our thoughts, emotions, and perceptions. There is no one else that could create your reality for you; so to blame the external world for your life's circumstances is to give away your power, but to take responsibility for your own life is to regain your strength and power.

I first encountered ayahuasca in December 2015 when I went to Iquitos, Peru, to research this powerful medicine as my undergraduate thesis project. I had been fascinated by the topic of ayahuasca and psychedelics ever since I took a second-year psychology class called "Altered States of Consciousness," taught by Dr. Imants Baruss. A few years later, I asked Dr. Baruss if he could supervise my thesis project in which I would go to the Amazon jungle to investigate whether drinking this powerful psychedelic could help people enhance their perception of meaningfulness in life. To most professors, this would have been an absurd request, but to Dr. Baruss, this was an opportunity to empower a student who really cared about this psychedelic medicine. So between planning the trip, battling with the ethics board to get the project approved, and mentally preparing for what was about to come, we finally saw my dream come to fruition by late 2015. Before I knew it, I was on my way to the Peruvian jungle with a backpack full of jungle gear and research materials. I spent one month at an ayahuasca center called Nihue Rao Centro Espiritual where I got to know the ayahuasca medicine through the Shipibo culture and where I collected data from 29 participants. Although I learned a lot from the research, I learned much more from having directly experienced the ayahuasca myself.

The first night I participated in ceremony, I walked into the *maloka* (the ceremonial hut) without a drop of fear, as curiosity and excitement overpowered any other emotion I could possibly feel. I couldn't yet understand how people could be so afraid of such a beautiful experience that I had read so much about. Well, it didn't take long for me to figure out why ayahuasca is so feared yet so revered at the same time when I finally tried the drink. After everyone went up to take their drink, the shamans blew out the candles and for the rest of the night we were left with the pitch-black darkness of the night and the lively noises of the jungle.

After 20 minutes of consuming the drink, I could feel the medicine slowly crawling up my spine, mildly resembling the sprouting of a wildflower. I smiled and lovingly welcomed the spirit of ayahuasca into my body with playful curiosity. Soon after, the sounds of the jungle started to get louder and louder, as if someone had cleaned out my ears and blasted my sensitivity to pick up on those miniscule sounds of the jungle that I could not possibly hear before. As the shamans started singing, I could feel many more sensations now coursing through my entire body; sensations I had never known before. At the same time, the visions were now flooding my perceptual vicinity. The more they sang, the more intense everything got. Louder, stronger, faster, and stranger. I was seeing things I could never have imagined, not even in my most bizarre dreams. All my senses were now wide open like a fully blossomed wildflower, as I could feel every little thing right down to the atomic pulsations. It didn't make a difference whether my eyes were open or closed, because in that state of mind, I could perceive so much more than my eyeballs alone ever could. The sacrament of ayahuasca had granted me direct access to this astounding world of energy. It was vastly big, playful, intimidating, eternal, and timeless.

When we use our five senses in a normal state of consciousness, we are actually receiving and interpreting energetic information, which the brain then translates into a coherent and whole perception. However, we receive very limited energetic information through the five senses, so we are mostly unaware of all the energies that are constantly influencing us when we are in a normal state of consciousness. When you drink ayahuasca the medicine cracks you open to a state of ultra-awareness and ultra-sensitivity so you can feel and sense energy like never before. It is like going from having only five senses to 500 senses within a matter of minutes. When this happens, you can see the energy around you come alive and your perception loses its coherence, as everything in this energetic world is interconnected and in a state of constant flux. Your perception also becomes much more chaotic and bizarre, as you can now interpret new forms of energy you could never interpret before. Everything becomes magnified and exaggerated, as the things that once felt good now feel extraordinarily good and the things that once felt bad now feel extraordinarily bad. So the ayahuasca experience is a double-edged sword: on one hand, it can be overwhelming, disorienting, and terrifying as the information overload can toss you around like a whirlpool that you cannot escape from. On the other hand, it can be extraordinarily empowering for it gives you a chance to look directly into the energetic world that dictates the physical reality we experience in everyday life.

Through my ayahuasca experience, I understood how everything in existence is a manifestation of energy. From the physical reality that we can see, smell, hear, taste, and touch, to the invisible planes of consciousness with all our thoughts, feelings, perceptions, and emotions. These different forms of energy carry unique information and are manifested based on their particular vibrational frequency. Under the influence of ayahuasca, you can see these energy forms represented symbolically according to their vibrational frequency. For example, the emotion of fear has a very different vibrational frequency from the emotion of love, so fear could be represented as devouring clouds of toxic smoke, whereas love could be represented as precious landscapes. I believe that the various visions that are commonly reported during ayahuasca experiences are mere visual representations of energetic frequency.

The job of the shaman is to look into a patient's vibrational field and interpret the symbolic representations of energy to make a diagnosis. The shaman then uses the vibrational frequencies of his/her voice to change the vibrational frequency of the patient and restore their well-being. I remember one night I went to the ceremony feeling quite sick from a stomachache, hoping that the medicine would heal me. When I drank the ayahuasca, I started to feel even more uncomfortable as I felt this sickness manifest into a dark dungeon full of spiders in the energetic world. Every sensation and emotion was now magnified and blown way out of proportion. When I was finally called up to receive a healing song from the shaman, I was so weak that I had to lay on the mat in front of him so he could sing to me. As he started to sing, I could feel his voice spread through me as divine light clearing out all darkness in my body. I could feel the vibrations so strongly, almost like a buzzing sensation. He projected the frequency of love to me, and as he did that, I visualized the light coming from my heart to spread throughout my body. The healing was a mutual effort between the shaman and myself, as he shifted my vibrational frequency with the frequency of his voice, but I had to be willing to align myself with that frequency to feel better. When I was finally brought back to my own vibrational state of well-being, I was amazed to have directly witnessed the inner workings of energetic healing firsthand.

When you have direct access to the energetic world with ayahuasca, you also see how the internal world of your consciousness and the external world of your surroundings are constantly influencing each other energetically. It is sometimes hard to tell what is coming from inside your consciousness to the outside world and what is coming from the outside world into your consciousness, as it really is just one eternal stream of energy, with no beginning and no end. It becomes evident that what you perceive is what you create, and what you create is what you perceive. What you receive in your experience is a result of what you offer to the world. If you let yourself be tossed into a feeling of panic during the ayahuasca experience from the overwhelming feelings, everything in your vicinity will respond to match the vibrational frequency of that feeling and give you an even more negative experience. The energetic shift is instant and unforgiving. If, on the other hand, you find yourself in a difficult experience and you actively use your consciousness to lift yourself up into a more peaceful state of mind, the energies around you shift to reward you with a more pleasant experience. I had to learn this principle the hard way, so I started using some anchors to help me through the ayahuasca experiences.

I call these techniques anchors because they kept me anchored and stabilized throughout the chaotic storms of ayahuasca. The first anchor I used was a deep breathing technique combined with non-judgmental awareness, which I had learned through the many years of my yoga practice. I would focus my attention on taking long slow breaths and this gave me the necessary strength and balance to keep me calm and relaxed when all I wanted to do was panic. The second anchor I used was to think about my mom and my dog whenever I would let myself slip in a state of panic. Every time I thought about them, my entire experience would be flooded with light, as they both represent what unconditional love means to me. These anchors are just a few examples of how you can use your mind to influence your state, rather than letting yourself be tossed around by the unpredictable and ever-changing energies within and around you. From this experience, I realized that the whole point of spiritual practices is to develop stillness from within so that the inner or outer turmoil of life could not easily rock your foundation.

Shamans strengthen and train their minds over a number of years to influence and shape-shift reality with their vibrational frequency. Although it may not be as evident and visible to us in a normal state of consciousness, we all have the power within us to do the same in everyday life. Ayahuasca helps us recognize this power within ourselves by increasing our sensitivity to energy, thereby revealing a deeper understanding of our consciousness in relation to energy. When you learn how to use your mind to influence the energies within and around you in a way that best serves you, you become the master of your fate rather than the victim of your circumstances. Not only can you rise above your negative emotions and dreadful circumstances, but you can also start creating the reality you want for yourself through the vibrational frequency that you offer to the universe. What you put out into the universe is reflected back to you, whether you are deliberate about it or not. Not being aware or deliberate about what you put out into the universe is to live life on autopilot.

It is the law of attraction that becomes so evident throughout the ayahuasca experience. Like gravity, it is my opinion this law never makes any exceptions and it never fails. But as vibrational beings with free will, we have the power to change the vibrational frequency of any manifested circumstance to best serve our needs and desires, just as shamans do. Ayahuasca ultimately shows us that we live in a participatory universe, and we get to choose how we want to participate in this universe through the will of our consciousness. So we can either continue living life on autopilot by letting ourselves get tossed around by the inner and outer turmoil of life and constantly blaming everyone else around us for our circumstances; or we can take responsibility for vibrational state by actively engaging our consciousness to create the reality that we want for ourselves and influencing those around us.

18

We've Come a Long, Long Way Together...

Freddie Findlay, Former Child Actor

Freddie Findlay was a critically acclaimed child actor, starring in films like A Feast at Midnight *and* Rasputin *before becoming addicted to alcohol and other substances at a young age. He came to Peru to find a cure for this intergenerational issue and ended up staying. He is currently an apprentice ayahuascero in Iquitos, Peru, in the heart of the Amazon Rainforest where he lives with his wife and two children. This is his story to recovery.*

∞

I have been living in Peru for the past 15 years and first started working with Ayahuasca over 12 years ago, and I have dedicated the past nine years of my life to working with the many sacred plants and medicines here in the Amazon. I have had the privilege of learning from many great "Maestros" over the years and I spent five and half years as apprentice to Maestro Roman Castillo Perez, and I run different retreats for individuals and groups in the Iquitos area. I'm also a founding member of the recently established "Ayahuasca Safety Association" which has been formed by many individuals, Peruvian and foreign, working with the various traditional medicines here in Peru to help create a higher level of safety and ethical practice for both practitioners and the many individuals that continue to arrive here.

Now is not the time or place to give an account of my life story, that would take entire book, but I will give a brief summary to give some sort of perspective…

I came to work with ayahuasca, "la medicina," and the plants here out of sheer desperation, and I can say I probably wouldn't be here today if it weren't for these sacred plants. I come from a good family and a fairly middle class upbringing in the west of England but my father suffered from severe alcoholism my whole life until his death in 2004, and I was abused and bullied at school when I was younger, and these experiences affected me greatly and I battled with various "mental health" issues, mainly depression, anxiety and addiction since my early teens. I was a hard drug addict on and off for almost 17 years and started using around the age of 14, spending most of my life addicted mainly to cocaine, crack and opiates.

Through my adolescence I was always in trouble with the law, and in and out of police cells and courtrooms, completing lengthy community service and probation orders for various drug-related offences, but luckily, I managed to stay out of prison. I spent years, on and off, in therapy and rehabs but nothing seemed to work for any lasting length of time. I came traveling around South America in 2005 to get away from life in London, where I had been living for the past five years, hoping that things might change. I had some amazing experiences during that time, but nothing changed really, and my drug use continued, and I dealt drugs to earn extra money as I traveled. I settled in Peru at the end of that trip, met my first wife and our son was born in 2007.

At this point I did have a few years when things calmed down after I drank ayahuasca for the first time, an experience that changed my life forever. I started working with the medicine as regularly as I could and began to slowly start working through some of my problems. Unfortunately, that was not to last though, and when I received a small inheritance I bought a bar, stopped the medicine work and fell back into using again. I had a lot of money at one point. I got involved in drug trafficking also, but after my behavior and using caused the breakdown of my first marriage I eventually lost everything and ended up smoking and shooting drugs again, penniless and suicidal, living on the streets of Lima for over a year. I was a broken mess and I needed help.

I knew I had to get to the jungle and the plants but I was stuck in Lima and flat broke. It was around this time that an old friend of mine, Stacy Povey (founder of DreamGlade), called me out of the blue and offered me a job managing his then business, Karma Cafe. I managed to borrow some money and a couple of days later I was on a plane headed for Iquitos. I arrived and started working with the medicine again. Shortly after I met Maestro Roman and started to heal myself with his help and I later left my job at the cafe to study with him. I still had a couple of relapses, but I managed to get through those times with the help of the plants. I also opened a small guesthouse with my wife, "La Casa Chacruna," and I started organizing and running retreats at a small center just outside Iquitos which I did for just over four years, and I finished my work with Roman not so long ago.

I continue to study "curanderismo" and holistic medicine with different teachers and still organize and run retreats in the Iquitos area, and I live a happy and content life with my family, free from addiction, depression and anxiety, having worked through what were my deeper issues.

I only mention the above to give an idea of what my life was like and how it is now in that it may give hope to those reading this struggling with similar problems. There is another way, and if you are willing the plants can help you find it.

Again, here is not the place or time to go into the many experiences I have had through using these amazing medicines, but I will share a little about how they have helped me personally.

Ayahuasca and the plants here have changed my life completely, and for me there is no denying their healing and transformational power and the immense all-round benefits that can be received from their proper use. I have drunk "la medicina" hundreds of times and had many profound mystical and deeply personal experiences. They grant us access to other dimensions and realms of existence where various different entities and spirits reside, and we are able to commune with them while there. They also give us access to our own subconscious minds and have a way of showing us exactly what we need at that moment.

Delving into my own subconscious, transcending my ego, and entering into a conscious dialogue at the same time, enabled me to work through my deepest fears and personal issues intrinsically in ways that years of therapy simply couldn't achieve. The resulting transformation I have gone through and continue to go through, what I have learned and continue to learn, have been truly liberating.

I would like to add at this point that only a few of the plants here are actually psychedelic in the traditional sense with rest being more psychoactive or psychotropic, Ayahuasca (brewed with Chacruna), Toé (Brugmansia or Datura which I will talk about later on), Magic mushrooms (Psilocybe Cubensis), San Pedro (Huachuma), Yopo (one medicine I have yet to work with), and I would also say Cannabis, being the notable exceptions. These have all helped me in ways that are simply astonishing to me even now but I would like to stress the value and power of dieting the various other teacher plants as they work in different ways and on many levels; physiologically, psychologically and spiritually, and the time spent in isolation while on "dieta" is invaluable and truly priceless.

Over the years I have experienced many things in ceremonies and dietas and been witness to many miraculous healings apart from my own. I will share a few experiences here which some may find incomprehensible or hard to believe but all that I say comes from my own personal experience and from assisting in the treatment of others.

The following accounts and subsequent healings were achieved through the use in most cases of a combination of ayahuasca, various medicinal plants, holistic practices and spiritual means as well as long periods to process and integrate for myself. To give some examples, I relived various moments from my life from both subjective and objective perspectives and I was also able to re-experience those emotions and feelings from before and work through those past experiences, make peace with myself and/or the people involved on a personal and spiritual level, accept and finally let them go: the grief, pain and anger experienced from my father's alcoholism and subsequent death; the guilt and sadness of not being there for my son through his early childhood because of my drug abuse; the shame I had of disappointing my mother and family; the remorse and guilt of what I put both my ex and current wife through as a result of my drug abuse; the anger and resentment towards those that abused and hurt me as a child; and the shame and guilt I had of all the wrongs that I had done in my past.

I was finally able to forgive myself and others, and let go of those things for good. I was also able to take a look at my own materialistic wants and desires and I now have a much simpler lifestyle, free for the most part from those unnecessary egotistical needs. I also understand now that all these things happened *for* me and not *to* me, and I am forever grateful for everything that happened to get me to the point I'm currently at. I still make mistakes, and when I do I try and find the lesson hidden within, to learn from and grow. There are always new challenges and things get difficult sometimes, but I try to watch myself constantly and not let my ego and emotions get the better of me. As result of those periods working with ayahuasca and various plants a fundamental change took place in me and I no longer have the desire to use drugs or feel depressed, and I feel immense peace and love for myself and others and I am forever grateful for that.

Also, the sacred knowledge and wisdom accessible and comprehensible through the use of these plants, combined with studying the various ancient teachings, has profoundly altered and expanded my consciousness and the way I see and understand the world and the universe, and all of this would not have been possible without the help of these sacred plants.

Just to make it clear, this didn't magically happen all at once. It was a long process for me personally, and it took years of "trial and error" so to speak, constant reevaluation, soul searching and many (some extremely difficult) ceremonies, "dietas" and personal work, and really this never stops but definitely gets easier.

Apart from my personal experience I have assisted and been witness to many other people's own healing and transformation over the years, many with serious personal issues and deep traumas: women that have been abused sexually and/or physically, some multiple times; people suffering with severe PTSD, many of those veterans; men and women suffering with severe addictions, depression and anxiety; people with "borderline personality disorder" and many other psychological and emotional disorders.

Also, many people with a variety of physical problems and illnesses, many caused by unresolved emotional issues, varying from neck and back problems, psoriasis, stomach and intestinal issues and also various viral and bacterial infections, eyesight problems, fungal infections, hepatitis and cancer. Most of these cases were treated not solely with ayahuasca but by a combination of plants, natural medicines and spiritual healing, and again is testament to the power of these plants and medicines and the knowledge and skill of the Maestros working with them, and I feel privileged to have been able to help in many of those cases.

I must also add that there have been many people that I have worked with over the years with no serious problems but after working with the plants have had their eyes opened somewhat and have found a way back to the spiritual in some way and that can only be good thing.

Iquitos is a truly unique place and has become the medicine "Mecca" with thousands of Westerners coming each year to drink ayahuasca and work with the plants, usually at one of the many retreat centers now in operation. Living here for the past six years I have seen and experienced a lot! There is no doubt to the amazing work that goes on here, but it has a very mixed energy and there are some things I think people need to be made aware of and look out for.

Iquitos is alive with people talking about their experiences with the medicine, healers and travelers alike sharing stories and swapping notes and some of those conversations can be extremely interesting and very enlightening from my own experience, but I do see a negative side also. I think it is great to share experiences and enter into constructive dialogue, but I also see a huge "expansion of egos" going on at the same time. It's an easy trap and one I myself fell into in the past, with a lot of "spiritual bypassing" and so-called "medicine junkies," who have some very fantastical and ungrounded ideas. When working with these medicines it is essential to stay grounded, take time to process and start putting things into practice slowly. The medicine is here to heal and teach us and expand our consciousness not our egos, and just remember that humility is the beginning of wisdom.

I also see many people coming into the city after retreats and drinking and sometimes using drugs. I would warn people against this strongly having made this mistake many times myself in the past, and I have seen what it can do to others also. The diet restrictions that are advised are not without good reason, and after a retreat people are extremely open energetically and the body needs time for the medicine(s) to settle properly. I have had to help more than a few people who became extremely sick after eating certain foods and/or drinking alcohol, and also some that quite frankly "lost it" after drinking and/or using drugs, and one friend of mine also died from an accidental overdose a few years ago.

Another important topic is Toé (Brugmansia/datura). I have had to help a few people that were given Toé, some knowingly and some not, at supposed retreat centers and by individuals in the city. Most Western-run centers do not use it in their medicine at all now, but a lot of local curanderos do use this plant, in small amounts, in their Ayahuasca, normally with no serious adverse effects. Given on its own this plant can induce severe psychosis when not administered by a trained curandero who sits with the person around the clock while under the effects. (I may add that it is an amazing medicine when used in the correct way.)

The people who unfortunately were given this ended up between realities, and in most cases, it took at least a month or longer to bring them back, and they self-harmed and did all sorts of insane things during this time. I advise people to be extra careful when drinking at smaller centers and with individual curanderos as some have been known to give people Toé instead of ayahuasca, usually to rob them, and it is used a lot for "brujeria" or black magic. There are also people selling liquor that contains the plant at the Belén Market. I strongly advise people to stay away from this plant in general. (Exceptions being those who may be studying the medicine and working with a trusted Maestro.)

I also know of some retreats in operation here run by people who have no real training or experience with the medicine, do not screen people properly, and sometimes use locals who can sing a few "icaros" to run their ceremonies but are not trained Maestro(a)s and this is extremely dangerous and irresponsible. "The road to hell is paved with good intentions," and one only need look at some of the accidents that have occurred, that could have been easily avoided, as drastic examples of what can go wrong. I would say this applies to many ceremonies being held in other parts of the world also. Unfortunately there are also many unscrupulous characters here and elsewhere looking to take advantage of people, especially when they are in an open and suggestible state, and I would just bear this in mind.

Thankfully there are many safe and trusted retreats and individuals working with the medicine both here and abroad with many new ones popping up all the time, and sometimes it can be a bit overwhelming when trying to choose the right place. On that note here are a few words of advice for anyone considering coming to work with the plants.

Do thorough research before choosing a retreat and don't just take the first spot in whichever center has space at that time. Check out the websites and reviews online, the blogs, and try and speak to someone who has been there before if possible. Contact the center by e-mail or telephone at the earliest opportunity and talk through your issues and queries. Check how many facilitators there are per group/person and some information on the curanderos they work with and how long they have been practicing, and also what plants they use in the medicine(s). Check each place you are considering has an adequate intake procedure and screening process/medical questionnaire that should cover physical and mental health history; if they don't then I would look elsewhere. Also check that they supply adequate diet requisites to be followed before, during and after your stay.

Check what medical facilities they have on site, how close the nearest clinic or medical post may be in case of an emergency and if they have qualified first aiders on site for your peace of mind. In the case of those that may be going to work with individual curandero(a)s and not at a known center then I would take extra caution and only go if having a recommendation from someone trusted who has been there or knows the curandero(a)(s) personally.

Have all financial agreements clear from the beginning and preferably get something in writing, even on a simple bit of paper. Also seek advice from someone with correct knowledge of correct dieting requisites and how to prepare beforehand, check that you are medically able to participate, and are not taking any medications or supplements that may interact with medicine, especially SSRIs/antidepressants and certain foods/ herbs, and do not have any existing mental health or physical conditions that may be noncompatible. Many local curanderos have very little knowledge of the possible interactions with Western medications and many physical and psychological disorders such as chronic high blood pressure or bipolar disorder for example. (Check with a GP with knowledge of ayahuasca if in doubt.)

People are waking up all over the world to the power of these ancient traditions and medicines that have been part of our existence and the evolution of human consciousness since before the beginning of recorded time. They have come back to us now at a very crucial moment, through divine intervention, to heal and teach us and get us back on the right path. It is our responsibility to seize this opportunity and collectively walk down that road together. They have become a global movement, and changing laws are making these powerful medicines more readily available and widely accepted, and although they have their base in those sacred traditions of the past, they have evolved into their current collective form and are in their relative infancy and we must nurture them with care, careful guidance and love if they are to have a lasting positive impact in the mainstream of today's social context.

The work being done here and elsewhere at the many reputable centers and by the many amazing healers and individuals working with these plants is truly changing and saving people's lives. And the thousands who have been healed and the many that continue to arrive seeking healing and spiritual enlightenment is testament enough, and I am grateful and honored to be able to play some small part in that. And there is no denying, when used in the proper way, the healing, transformative and consciousness expanding power of these sacred "psychedelic" medicines.

19

That Jungle Magik

Sabrina Pilet-Jones

Sabrina Pilet-Jones is an urban farmer, floral artist, witch, integration coach, and founder of Sabrina's Garden, through which she offers locally grown herbs, flower arrangements and botanical jewelry to reconnect urbanites with the beauty of the natural-scapes around them. Sabrina, the assistant farm manager for the Urban Farming Institute, works with and advocates for plant medicine education and harm reduction and leads various creative projects throughout Boston, where she lives. In 2020, she co-taught "Juju Box: Sacred Plant Rituals of the African Diaspora" at Boston Herbstalk with her partner, Yoruba Practitioner Arirá Adééké.

Sabrina began practicing Wicca at age nine and has explored spiritual/ magical studies, including hoodoo and shamanism, throughout her life. Studying African American spirituality and ancestral universal laws of living in harmony with Dr. Nteri Nelson empowered Sabrina to continue her own personal spiritual exploration and help others heal. She has received several Cosmic Sister awards and grants, including an immersive ayahuasca Plant Spirit Grant, a Women of the Psychedelic Renaissance Grant, an Emerging Voices Award, and an Ancestor Medicine Award.

∞

I had knots in my stomach, an anticipation I had never felt before, an excitement for adventure I already wanted more of—and the journey was just beginning. I never in my 29 years imagined I would find myself boarding a plane alone to South America to meet with 22 strangers and trekking through the Amazon jungle to try what is known as the shaman's brew, ayahuasca. The journey itself propelled me outside of my comfort zone to reach down deep for emotions and energies I had never felt before. I had to rely on myself, sit with myself, and in ceremony, confront my lower self.

My living accommodations, a *tambo* hut nestled within the jungle, had a small, cute porch looking out to water that was home to the biggest fish I've ever seen in person. I have a standoff-ish relationship with bugs, but I felt like they were calling out with messages relevant to my journey. I made acquaintances with the lizard that greeted me each morning by flopping from the ceiling down to the floor with a splat, just as the rooster began its morning song. The insects have many lessons to teach even if they freak us out, and I learned quickly that they cannot be avoided in the jungle.

My first night there, I was greeted by an Amazon (re: gigantic) cockroach that hitched a ride on my purse, which I noticed as I went to put the bag on over my neck. I had to work through that one for about an hour. One yoga session later, I was walking back to my tambo ready for sleep yet praying the hitchhiking bug had found a different place for the night. Not only were my prayers unanswered, but some friends had joined him on my dresser, and they were feasting on the beautiful feather earrings I had purchased in Iquitos.

I dove under my mosquito net, tucking it in tighter than a facelift, and laid down wishing I could sleep with my eyes open. The nighttime jungle symphony both lulled me to sleep and jolted me awake throughout the night with a mix of soothing outdoor sounds, rambunctious monkeys playing games in the trees, and a huge fish plopping into the nearby water like clockwork—an interesting balance of *this is amazing* and *this is annoying*.

My days were filled with learning about plant remedies and the traditional plants that the shamans worked with throughout the jungle, permaculture jungle walks, floral baths, bonding moments, and a shitload of spiritual inner work and reflection. Delving deeper into the world of plants and plant spirits, I felt steeped in ancient wisdom and intrigued by the magikal energy the jungle illuminated. I wanted to hold this energy with me forever.

Most of my nights were filled with ceremonies that propelled me into otherworldly dimensions, memories, and ancestral meetings. Starting at 8 p.m. sharp, we would gather in the maloka (a round ceremonial space) on mats arranged in a large circle, our makeshift travel mechanisms for the night along with the tools we would need: a bowl to catch vomit, a plant-based water made of rue to help ward off any negative energies we might experience through our journey, and *mapacho* (sacred Amazonian tobacco used for spiritual protection). The shamans would enter and take their places in the center of the circle, clouds of *mapacho* smoke outlining them like chalk dust as natural wisps of moonlight crept in. We would go up one by one to drink, a powerful and scary experience. Knowing that you're making the sound decision to ingest something that will alter your consciousness in a way you can't comprehend fills your body with butterflies. Partaking in ancient herbal medicine and rituals strikes a fire inside that is inexplicable and incomparable.

On the first night, I walked back to my mat and waited, scanning the room with low eyes to see how my fellow passengers were making out. Some people seemed to be feeling the effects as the night kicked off to the tune of coughing, spitting, laughing, and farting. My body began to relax a little bit and feel heavy, but that didn't last long.

The shamans began to vocalize a beautiful song. Each of their voices was like a different chord to one instrument. My vision began to blur, and my body began to buzz as the plant medicine started to take over. I was suddenly surrounded by what felt like a million eyes looking at me with curiosity. My vision was taken over with vivid patterns and colors, and creatures played peekaboo with my peripheral vision, jumping in and out of my sight, peeking in every now and then to look at me. I sat back as a spectator, trying to make sense of all that I was seeing and being shown, yet it all appeared to be in a language that was foreign to me—the language of the plants.

The shamans moved between us, singing to each of us individually, and their songs penetrated through to my energy body and rebalanced me from the inside out. My body seesawed between feelings of warming comfort and empowerment to extreme discomfort and fear, and that night, I learned that both feelings are necessary for growth. I reached for my herbal rue cologne whenever I felt the need to tether my consciousness back to something familiar—which was often.

I Am the Daughter

I was raised by my second cousin and her partner. One of my sisters told me who my biological mother was when I was three, but I didn't know my biological father and his family until I was about 16. My adolescence was steeped in identity issues and low self-esteem, which led to rebellious, self-harming behavior and, later in life, depression. When I began to explore what my adoption meant to me, how it shaped who I was, my thought processes, and most importantly, how I viewed myself in this world, I realized I was confused about my place in this earth and how I came to be.

I went into ceremony plagued with guilt because I wanted to shelter my adoptive parents while I was on the journey to reconnect with my biological parents and families. This inner split feeling is something I may always have to deal with. Even writing about my adoption has tweaked that inner sway between low-key fear of, *will this hurt my adopted mom?* and *screw this, I need to tell my story.*

During ceremony, three of my grandmothers came in a powerful moment to show me my place in my family. I heard a voice that night before they arrived, unrecognizable but clear, stating firmly, "Daughter of Susan and Drake." Things began to move and shift around in the room, and I felt a deep inner whole feeling I had never felt before. This one simple statement shifted my perspective, and I believe it was the beginning of the shift in my healing and knowing my place. I have always viewed myself solely as the daughter of Barbara and Thomas, my adopted parents. I had never even said my name and my biological parents' names together in the same sentence. Hearing that, I felt like I shifted into my rightful place. I felt confident in owning these two truths. I am the daughter of Susan and Drake. I am the daughter of Barbara and Thomas.

Of the three grandmothers, two were my adopted grandmothers (my great aunt and the mother of my adopted father) and one was my biological grandmother (my bio father's mother). I did not feel an ounce of fear when they arrived. I felt comforted. My adopted grandmothers came to me first, bringing strong memories of my childhood that drew me closer to them, feelings of summertime sun, the smell of warm fudge, and old-school couches pressed against my face. I missed them deeply in that moment, but I knew they were with me. I felt overtaken with pride, comfort, and unconditional love. When my biological grandmother appeared between my other grandmothers, she was larger, majestic. I felt a connection to her like I never had before, this grandmother I unfortunately didn't get to know well, who had passed away the year before I went into ceremony. She arrived with a fierce, powerful energy of connection, ownership, and remembrance, a proud black woman, solid in who she was.

I felt her as a missing link in my life. At her funeral, I hadn't felt like I had the right to cry. I'd been deprived of a connection with the woman my siblings said was one of the best grandmas, which was especially painful after I learned I had passed by her often while she was a crossing guard in an area near me when I was a child. Now she stood before me in ceremony and said, "You are of me." I felt that to my core existence. I felt like I was sitting at the foot of an abnormally large elder, reminding me to remember where I came from and from whom I came. She turned and acknowledged my adopted grandmothers, saying, "Thank you for being in her life while I could not be." My adopted grandmothers spoke to me. "You need to find out who you are and connect with your biological family," they said. "Our children will be OK. You cannot fear hurting them. Seek your truth."

The Rearranging

The second night was the hardest night for me, and it was the night I took the smallest dose. That night the medicine came over me fast. I could feel myself spiraling out before the shamans even began singing.

Was this *supposed* to happen? I had learned the night before not to fight the medicine with body or mind but to allow the plants to show what they wanted to show me. I struggled to stay sitting upright but soon decided I couldn't, that I had to lay down. The medicine felt as if it were creeping up my body, rearranging things. My body felt like it was swaying and on water, then I was aboard an old-style canoe with a straw canopy, wrapped and lying on the floor of the boat while faceless beings with straw hats and lanterns worked on my body and guided the boat out into the water. They seemed connected to the shamans' songs and to the water they were guiding me over. I felt them touching, scanning, and probing my body, then they brought me to a lab-like examining space where faceless, ghostlike beings performed energetic surgeries over my body. They used lanterns to check over me, their movements mechanical with an ancient mysterious undertone to them. The beings moved in and out as if in a hospital rotation, probing and poking like inquisitive doctors. I could hear noises that sounded electronic. I was afraid yet felt safe, like I was being examined, rearranged, and healed at the same time. I lay there at the whim of the plant spirits, my body experiencing so much at once and my mind struggling to make sense of it all.

I felt a pressure building up, a contracting energy moving its way up my body. I was in total discomfort, seeing spiders, and I couldn't tell whether it was in my mind or if there actually were bugs on me. "I am finding comfort in the discomfort," I began repeating to myself.

I laid on my mat and reached down deep inside of me for comfort, to an inner space of peace that exists inside of me at all times, no matter what. I had to *be that comfort* for myself in the midst of the chaos I was experiencing. It felt almost like I removed myself from the experience and became pure awareness, which always exists and never owns, but just is. I had to not own what was happening to me and separate myself to get by. I had to be my own comfort. This struggle seemed to last for hours.

I felt the pressure break through my head like a busted pipe, and I purged. I struggled to place mundane facts in my head to gain grounding. "Is it Tuesday? It's Tuesday, right? What time is it? I'm in Peru in the jungle. Am I dying? No, I'm alive." I questioned and consoled myself while trying to capture my bearings. The room was vibrating with so many sounds, and the shamans' voices were a constant familiar noise I could adhere to. I began to feel a clarity come over me. My nose opened up, and my ears and head. I felt as if air could move through me, totally open in a way I never had been before. It was as if my pores opened up to the widest they could go; I felt connected and open to everything, aware of myself and how the space felt around me.

The next morning my body felt as if though it had been pushed through the ringer. I missed herbal remedies and even breakfast. I was exhausted, but somewhere deep inside, my spirit felt empowered. I had found comfort and empowerment within discomfort.

I Want My Body

The elephant in the room on this spiritual trip was my body dysmorphia. Every time I sat in ceremony, I wondered when this obstacle would reveal itself. I had a preconceived idea of how the plant would help me heal through it, but you cannot predict the way the plants will convey their messages to you.

After the second ceremony, I was apprehensive. Was I *reeeaallyyyyy* going to put myself through that again? Was it all worth it? Going all the way to the jungle just to learn how to love myself? How foolish am I?

I drank more to find out. This time, I sat up straight, determined to control my body functions. My hands were sweating in anticipation as I waited for the songs to begin. I felt my stomach begin to flutter and my eyes get a little heavy. I closed my eyes and swayed and laughed. The *mapacho* smell crept into my nose, and I saw the shamans clearly while my eyes were closed, illuminated with and surrounded by blue light, blending with various animals, half-human, half-animal beings—a very beautiful image to see.

I waited for the medicine to take off, but it didn't, so I was fully present through each of the six *icaros*. I looked around and watched everyone and wondered why tonight not much was happening to me. I laid down and attempted to relax but couldn't. I laid in a state between relaxing and what the heck, but not "tripping." My body couldn't find its place. I moved between a repetitive loop of sitting up and laying down. Maybe the brew was starting to hit me after all. I felt weird and hot and nauseous, and I thought I could purge for relief.

I got up and realized I was a little more unsteady than I thought. I made it to the bathroom with help. I sat down and looked at my feet and hands, weirdly unrecognizable as I sat on the toilet, sweating. I took off my clothes and showered, which didn't feel as good as I thought it would, and then I was cold and in the bathroom with bugs, semi-tripping, and I had to figure out how to get my clothes back on, which I did. After a standoff with a huge bug, I decided it was time to get out of the bathroom and return to my mat.

Back in the ceremony, I tried to relax into things, but I was still tense. I could feel the aya kicking up. I laid down and tried to brace for what was to come. I saw random lights sparkling. I looked over at one of the shamans and saw a blue grid pattern coming from inside of her. The jungle outside seemed to glow supernaturally, and everything looked and felt alive. The rain was pounding outside, and I sought solace in the sound of water. Lying on my mat, I felt as if beings were beginning to come around me, except this time I didn't feel the same curious probing energy as the previous night. I heard the beings chattering amongst themselves but couldn't fully hear what they were saying.

I felt fear rising. I heard voices again.

"Her body?"

"Yes, her body."

"She doesn't want it."

"Let's take it, then."

Numerous hands began to grab at my body. I kicked my feet and hands and felt words forming deep in my throat, words I'd never said or owned inside of myself.

"I want my body! It is mine, and I want it!"

The hands stopped grabbing. I cried and hugged my body and rocked. I thanked my feet and my hands, and I thanked my body, the body I had verbally abused and at one point in my life even physically abused. This body I had hated and yearned to change. Nobody had ever tried to take it before.

I Am My Ancestors

The final ceremony sealed my work with the plants and allowed me to tap into the essence of all that I could be. Feeling comfortable and empowered, surrounded by my ancestors, I cried and felt as if I were at the feet of all those who came before me. I sat up, feeling myself morphing into hundreds of different images per second, as if I were scanning through a rolodex of ancestors. I felt a glimpse of who I was, empowerment through lineage. My magic is based on my lineage, and tapping into that will help me complete my ancestral work.

This journey has taught me so many things. Some things are hard to explain, and some things I hold as a constant reminder. The entire experience changed me forever as a woman—it was a personal rite of passage. I pushed past my comfort zone in so many ways, and every single experience was necessary for growth.

I want to keep this magik I found within the jungle with me forever. I try to find tiny pieces of it in my everyday life to keep me going. Ayahuasca is not a magikal pill. It's hard, deep, transformative shamanic work that forces you into the deepest, darkest parts of your self to find the light—the *unique* light we all hold. I left with a strong desire to expand my connection with plants, to continue my research into indigenous plant remedies and now psychedelic plants for journeying. Most importantly, I returned home with a new perception of myself, and so many other things, and I brought home the pieces of medicine that Mama Aya gave me, so I can get by when the magik of life isn't as easily visible.

20

How I Healed from Trauma, Addiction, and Depression Using Psychedelics, Yoga, Meditation, and Deep Psychological Inner Work

Laura Matsue

Laura Matsue's journey began when she went through several "dark nights of the soul." After struggling with addiction and decades of trauma, she had a profound awakening experience in 2012, which led her to the spiritual path. This "awakening" brought her to daily yoga and meditation practice, which led her to travel the world in search of spiritual knowledge to study alternative healing methods.

Since then, Laura has studied various modalities between Western psychology and Eastern spirituality. Her main interests are psychospiritual therapies; she has studied Jungian psychology, psychosynthesis, and various somatic therapeutic modalities. She is also a Tibetan Buddhism student and an avid meditator and yogi.

Laura is a somatic counsellor and a practitioner of Compassionate Inquiry (Gabor Maté's modality), Internal family systems therapy, and practices other somatic and psychospiritual therapies in an embodied and trauma-informed approach. She is also a teacher of yoga and meditation and an evolutionary astrologer. She calls this "psychospiritual coaching" as she works with the whole body/mind, soul, and spirit.

She has hosted courses, retreats, and workshops with her husband Bernhard. They currently host a psychospiritual group program a few times a year called Embodied Soul Awakening and a host a podcast together called Cosmic Matrix.

∞

The following is a story of how I healed myself from trauma, addiction, and depression using psychedelics, yoga, meditation, and deep psychological work. Although psychedelics have been pivotal in my journey, through my own trials, I've realized a holistic healing model is the most beneficial for lasting change, which has become central to the work I offer now as a psychospiritual coach and somatic counsellor.

My healing journey began in early December 2012. I decided that was the day that I was going to kill myself. Since I am writing this now—clearly, that plan didn't work out. A cop intervened when I was going to buy the drugs that I hoped would kill me, which gave me enough time to reconsider my decision.

Although I didn't know it then, this moment was a turning point in my life and was the beginning of my decision to take my healing into my own hands. This "dark night of the soul" would eventually turn my heart towards God. Psychedelics played a part in that healing journey, along with somatic, psychospiritual therapy, yoga, meditation, and inner work.

The drugs I was on at the time (suboxone and methadone) made me feel a lot worse. I felt hopeless from my visits to addiction specialists only to have them offer these as the best pharmaceutical solution. These drugs made me more depressed than ever, physically weak, unwell, and emotionally numb. While they got me off the other painkillers I was using, they weren't providing the deeper healing my soul sought.

After the suicide attempt, things turned from bad to worse. Sometimes, I've found that you need to spiral down to the bottom before your soul reaches toward the light. I got together with an old boyfriend who was violent, and angry and seemed to live in a world where his version of "love" meant abuse and control. Truthfully, I wanted someone to "save" me. After one terrifying encounter with him where his fist was inches away from pummeling me in the face, I realized that my life was at risk living with him. That night, feeling like I had nowhere else to turn, I prayed to God. I had no money, no friends in the city I had just moved to, and no job.

Being a relatively non-spiritual person back then, I remember feeling foolish at the time, getting down on my knees and praying to a God I wasn't even sure existed. Still, I remember asking with the most sincere part of my heart: "Dear God, if you can get me out of this situation—I will do *anything*." Shortly after I got up, someone I hadn't heard from in years texted me and offered me a place to stay. He had no idea of the dire situation I was in, and he was shocked to hear my story when I arrived.

This moment was a major turning point in my life that began a long journey of healing. From that moment, I decided that I was going to live.

I eventually moved into a shared apartment in another city. I began a long journey of rebuilding my health in body, mind, and spirit. I started with a daily at-home yoga practice, something I would've never imagined myself doing a few years ago. During these practices, I often cried; frozen grief from decades of trauma came out of my body while doing certain deeper asana poses. I remember thinking in these moments, "Wow, my body seems to be holding onto memories from the past few years."

I was still having symptoms of depression. I constantly focused on the negative and felt deep self-loathing. I realized that I needed to change something, and I remembered a research paper about ketamine I found years earlier and how it could cure "treatment-resistant depression."

I recalled reading about how it specifically affected the NDMA receptor in the brain, which is partially responsible for linking present-moment experiences to the past. I realized the trauma I had been living through over the past few decades (pretty much from early childhood) was my past being carried through to the present.

I felt I needed something akin to a "brain reset." I was unable to experience joy or pleasure anymore. So, I read up on ketamine, and I decided it would be worth trying to get out of this intense depression, so I could at least be functional at a job and get my life together again.

I was able to find someone in town who supplied ketamine. I have to admit, back then, ketamine treatment centers were rare, and I was taking a big risk by taking drugs from an unknown source that could have seriously harmed me. I set up a sacred space in my room to do my own "ketamine therapy." I was planning on putting soft, spiritual music and Sanskrit mantras and went into the K-hole. My journey was unremarkable, but I remember hearing the sounds of ringing bells, also known as "the sounds of the universe," while entering that space.

What happened after, though, was life-changing. I remember walking down the street on my way to a new job I had just gotten hired for and suddenly feeling alive again in a way that I hadn't felt in many years. There was a freshness to life. The world was colorful and friendly instead of scary and hostile. My inner child felt happy again—I felt joy for the first time in years. I could laugh again. I knew this wouldn't last forever, but this relief gave me enough of an "open window" (of about three months) to get my life back on track. I started working out every day, focusing on mastering nutrition and food as medicine, and slowly but surely, I was able to get my life back together, piece by piece. I felt like I had been given a new lease on life. I was reborn.

Unfortunately, this rebirth came with a caveat. I wasn't "cured" after that one session. My relationships were still a mess, and I still experienced PTSD flashbacks in certain situations that reminded me of some of the worst traumas of my past. These flashbacks put me into disassociation and extreme panic when they hit, and they were very confusing at the time.

There was way more healing to do, but now, I felt capable and ready for the task.

About a year after this experience, while working at a catering job at a castle, I ran into a soft-spoken guy with long hair. He was one of the first people I met during that period of my life who was genuinely nice to me. He was big into doing "ceremonies," and he always talked about this psychedelic plant called Ayahuasca.

This was the first time that I heard about that plant. He told me not to seek it out, and that if it were appropriate for me to work with her, *she* would find *me*. I took that statement to heart but never forgot about this mysterious plant being called ayahuasca.

Fast forward to a year later, I'm working synchronistically at another castle on a film set. On that set was a stunt woman, originally from Detroit, who now lives in LA, who went through a near-death experience that led her to undergo shamanic training. She now serves ayahuasca. She was strong, bold, and powerful, and seemed to embody what I knew about the ayahuasca spirit.

I flew to LA months later to be at one of her ceremonies. I brought blankets, crystals, and psychic protection items. I was years into my daily yoga and meditation practice at the time. I wasn't sure what would happen, but I felt safe with her leading the experience.

When we drank the medicine, I remember it kicking in reasonably quickly for me.

The first thing I remember being shown was the depth of the trauma I experienced, especially the sexual trauma. This was the first moment I realized how much trauma I was carrying and how painful it was for me. I had shut down/suppressed a large amount of it to survive.

I sobbed for almost the entire ceremony witnessing the abuse I experienced. It was the first time I truly felt what it was like to go through that. I grieved for all the pain I had experienced, which I had to shut down to survive. I grieved for the lost little girl who found drugs as the only way to soothe this trauma, only to end up worse than before. This personal grieving turned into something transpersonal; it became grieving for the damage done to the feminine and the earth. I realized that my pain was just a microcosm of that macrocosm of how we have ravaged the earth without any care or consideration for the impact. During what seemed like hours of sobbing, an archetypal spiritual figure eventually came in and held me with the deepest, most profound maternal, nurturing love I had ever felt. It was the presence of every grandmother figure I have ever known, which then morphed into my own loving, caring, biological grandmother. This Divine Mother figure held me through this experience with profound compassion.

This was a turning point for me. I uncovered decades of trauma I had suppressed in a single ceremony. Finally, I emotionally felt some of the pain deeply hidden behind the walls of my psyche's defensive mechanisms and I was able to see the weight of the trauma that I was carrying. This grieving gave me space to begin healing that trauma.

After this initial ceremony, I wanted to explore ayahuasca further. I decided to go deeper and visit the Sacred Valley of Peru, which is basically like a hippie epicenter for all sorts of shamanic healing.

I did several more ceremonies in Peru, but they were admittedly way less potent than that initial session.

I also witnessed in the Ayahuasca scene in Peru what I would call "spiritual bypassing," John Welwood's term for when we use spiritual ideas to bypass critical psychological developmental stages. Many people down there seemed to take on new spiritual identities, and were quite emotionally immature, including me.

This is when I came across shadow work, Carl Jung's theory for working with unconscious parts of ourselves that we suppress. While in Peru, I realized that these ceremonies were helping me less than I had hoped. I had to do more. So, I began to read books about depth psychology.

My first big moment was when I came across a book on shadow work. I realized that, like these people I was seeing, I was also taking on a new "spiritual" identity in an attempt to bypass facing past trauma.

Doing shadow work allowed me to accept the unpleasant parts of my past, especially the shame I carried from my past opiate addiction. I finally started to feel "whole" again by accepting the parts of myself that I tried to deny and leave behind. Through shadow work, I realized that I didn't have to hide elements of my past anymore—they were part of who I was.

This experience in Peru was another turning point for me. I realized that plant medicine, yoga, and meditation alone would not heal me. Instead, I needed to engage in deep psychological inner work.

After this experience, I moved into an Ecovillage, an intentional community on Vancouver Island off the coast of BC. I had spent a few years isolated in my healing and realized I needed to be around other people. There, I rediscovered the healing power of community, nature, working with the land, and just being with heart-centered people. It was a type of communal repair for me, welcoming me back into the world.

While living there, I also deepened my spiritual practice. I went away for my first 10-day silent meditation retreat. I loved the silence and space to be with my body and mind. I wasn't expecting profound existential sadness to come over me about halfway through the retreat. I realized how painfully alone I felt and how desperately I wanted a loving relationship to heal the pain of my past.

On the eighth day, something broke. I had a profound experience of grace, feeling true peace and love while looking out the window, watching the snow gently falling on the ground outside the retreat center. It was then I realized—that this was what I was looking for: a relationship with the Divine. There was still a part of me that believed I could only be happy in a relationship with another person. I realized that this love I sought was available in a relationship with God. This led me to deepen my meditation practice, which still is, to this day, an anchoring force in my life.

A couple of months later, I traveled to India to further my spiritual practice and take yoga teacher training. I meditated for two hours a day at various temples and ashrams, and sometimes did up to four hours of yoga a day. In between, I read various books on psychological healing and trauma.

The profound experiences I had in India were too many to count. I hiked to the world's highest Shiva temple, went to satsangs with Prem Baba, a guru who offered talks there, and meditated in an infamous cave where many saints had had enlightenment experiences.

One of my most profound experiences was at the ashram of Prem Baba's Guru.

Earlier that day, before visiting the ashram for my morning meditation, I ran into some wandering sadhus, ascetics, who offered me some hash. They were wild, technically homeless, had dreadlocks and survived on very few material possessions. Though I tried to stay sober in India, I smoked with them and then went to the ashram, even though there is a blanket rule in all ashrams against using substances.

I went towards the meditation room, feeling profound shame and guilt for what I'd done. I couldn't bear to bring myself into the meditation room. I thought I wasn't worthy of entering because I had smoked hash earlier that day.

Instead, I listened to the Gayatri mantra outside the meditation room on my headphones, meditating and singing along. The guilt and shame came over me so strongly. Then,

I felt something so profound that it is hard to explain in words. I had a transcendent experience where I realized that even my most profound suffering had led me to deep wisdom, and that the constructs of "good" and "bad" were merely judgments of my own mind. The presence of the Guru entered my consciousness, and instead of shaming and guilting me for breaking the ashram rules, I felt his endless compassion and love. I could not stop sobbing while experiencing this level of love, experiencing what I would call "divine grace." It was a feeling of being instantly forgiven for anything I thought was "wrong" with me—especially around my guilt/shame for using psychedelics. I realized my guilt and shame were self-made, and it was not some judgment from God. It was the most profound experience I've had in my life, to date.

That experience has become a beacon of light for the rest of my life and still is to this day.

The only quote I've seen that captures some of what I experienced that day is this:

Once you experience being loved when you are unworthy, being forgiven when you did something wrong, that moves you into non-dual thinking. You move from what I call meritocracy, quid pro quo thinking, to the huge ocean of grace, where you stop counting or calculating.
– Richard Rohr

I had experienced this Divine Grace at the moment I felt the most unworthy. This experience made me realize that this grace was everywhere; even in my darkest moments, even when I felt like an absolute "sinner," there was a hidden hand guiding and loving me all along.

Fast forward to today, almost six years later. I am working as a psychospiritual coach, weaving together depth psychology and spirituality and working with groups of people, guiding them through programs of yoga, meditation, psychological work, and spiritual work. My personal healing experiences and professional training in various modalities of somatic therapy and psychospiritual therapy (Holistic Counselling, Compassionate Inquiry, Internal Family Systems Therapy, Evolutionary Astrology, Psychosynthesis, and as a yoga and meditation teacher) have all informed the work I do.

Through my own journey, I've realized that since humans are multidimensional beings, our healing needs to be holistic and multidimensional too. It wasn't one single thing that helped me change and feel better; it was a multitude of things, yoga, meditation, nutrition, psychological work, somatic work, psychospiritual work, psychedelics, and my connection to God, which all got me from that dark night of the soul I was in, to the beautiful life I share with my husband and dog now.

Watching some of the treatments I used become available therapeutically now, I feel that humanity is at a breakthrough in healing. Even five years ago, when I started this journey, new therapies and healing methodologies came to the forefront, which weren't as available back then.

And like my journey, I feel that true healing has to address the whole human being: in mind, body, soul and spirit—bringing us back into our bodies, awakening our souls, and turning us back towards the most important relationship of our life—the one we have with God.

21

The One True Law

Elizabeth Bast, Author of *Heart Medicine: A True Love Story*, Iboga Initiate

Elizabeth Bast serves as a writer, certified yoga teacher, performance artist, Bwiti initiate and traditionally trained ceremonial facilitator, speaker/educator, and holistic coach specializing in sacred plant medicine support, spirit-led addiction recovery, and visionary life design. She is a lifelong devotee of plant medicine, ceremony, and prayer, having been raised within a long family line of plant healers and an inter-tribal community with traditional ceremony. Her greater intentions are to help facilitate 100% natural bliss, synergy between culture and nature, and conversations between the old and new ways of knowing. She studied at New College of San Francisco with an emphasis on Art and Social Change. Bast completed the transformational coach training program with Being True to You.

Bast is the author of Heart Medicine: A True Love Story, *an intimate memoir about a healing experience with the African sacred medicine, iboga.*

Bast is a recipient of a "Women of the Psychedelic Renaissance" grant from the psychedelic feminist organization Cosmic Sister and a member of Cosmic Sister's Expert Advisory Circle.

∞

Long ago, when the second woman ever in the history of the world ingested iboga in the deep jungle of Central West Africa, the spirit of iboga appeared to her and introduced itself. "Hello. I am the spirit of iboga. I have been watching you human beings for a long time. I have been listening to your questions... And I have come to answer your questions."

We human beings are curious creatures, with many cogs and wheels in our heads. We had been asking many powerful questions such as, "What is the meaning of life? What is God? Where do we come from?"

When I hear, "I have been watching you human beings for a long time," I sense that iboga had been watching us develop for eons, since we were simple lifeforms, like a curious parent watches a child grow up.

Co-evolution is the greatest love story in the world. The iboga medicine engineered itself to help us, to teach us about the power of our minds, fit into our receptor sites, cleanse our bodies, and reveal to us the ineffable answers that can incite respect for our own life and the lives of other beings, if we are willing to come in a good way. Why? *Just because it can.* Because nature is opulent—and Creator loves to admire itself through our eyes as we journey through the game of embodied awakening.

The iboga medicine has many functions—in micro, macro, and moderate doses—including spiritual initiation, diagnosis, prescription, communal ceremony, hunting, and healing. It is highly intelligent, a polymath of plants. Among these purposes, iboga is also known to help a practitioner see into the future, not one that is set in stone, but one of various doorways and crossroads that we can access through our choices. If iboga can help someone glimpse into future possibilities, it must be able to do the same for its own purposes. And this ancient plant has far-reaching vision.

Iboga must have seen the potential tragic trajectories available to the magnificent biotech machinery of the human being brain... and all the soul sickness and suffering that can arise from our unanswered questions. It's amazing *just how much destruction* can come out of a question like, "Why did this happen to me?"

Suffering humans means a sick (or dying) earth. Globally, there is an estimated minimum of 190,900 premature deaths caused by drugs (range: 115,900 to 230,100). Opioids account for the majority of drug-related deaths, and in most cases such deaths are avoidable.[1] North America continues to experience the highest drug-related mortality rate in the world, accounting for 1-in-4 drug-related deaths globally.[2] More, the pollution being released into the water from SSRIs like Prozac and hormone-altering pharmaceutical drugs are causing fish, birds, and other creatures to reduce mating, feeding, and predator avoidance behaviors that are essential for survival and procreation.[3,4,5]

Iboga must have predicted our potential for ecological destruction—which begins, guess where, in the human mind. This is not to say that it's as simple as just changing our minds. Suffering is complex and multifaceted. It's going to take a whole worldwide-tribe overhaul. And we need all the human-power imaginable to do that. However, our suffering, as we know it, being a biopsychosocial phenomenon.

"I love my life," said my husband, artist Chor Boogie, the day after his first healing ceremony with the iboga medicine, an experience that freed him from the slavery of a heroin relapse. Along with a profound physiological detox, this intimate knowledge about the preciousness of life is one essential element that can help to heal the very roots of addiction—and self-destruction. "I never want to disrespect myself again," he added. "Wow, the earth is so beautiful," he swooned, looking over the land with his polished eyes, as if for the first time.

There is only one law in the Bwiti: If you abuse nature, the price is misery. And the catch is: *you* are nature, too.

All entheogenic plant medicines can help to dispel the grand illusion of separation and teach us about our interconnectedness with all of nature. We are a *part* of an ecosystem, not separate from it, and certainly not superior to it. We are a part of the great body of Gaia. With our big fancy brains, humans can sometimes forget that what poisons one will eventually poison all.

Iboga knew: Human being pathologies could eventually one day threaten life and the ecosystem as we know it. And human being wellness could not only create harmony between human culture and nature, it could create *synergy*.

The sacred medicines have been slowly gearing up for this moment since we first crawled out from the ocean's womb onto dry land. The medicines are on the move now, say my African friends, from the remote jungles into the metropolitan hubs of humanity. Iboga was hidden for a time, and now it is revealed for a reason. The master plants have their own agenda: Life loves for life to continue—all while we come to realize that all of life is born of the Divine. May we hold these medicines with great care and respect as we receive their gifts.

These expressions arise from my personal relationship with the iboga medicine. These statements do not come from any dogma or adopted belief. Beliefs are known to be dangerous in the Bwiti tradition. Knowledge, however, is something different from a belief, and it is always confirmed from direct experience, not hearsay. At its core, the Bwiti is simply the study of life.

BASSE! (Truth. Yes. So it is. I agree.)

This segment of the origin story of iboga was passed on to me within the Missoko Bwiti tradition from Gabon, and variations can exist within different lineages.

References

1. The United Nations Office on Drugs and Crime (UNODC) "2017 World Drug Report": http://www.unodc.org/wdr2017/field/Booklet_1_EXSUM.pdf

2. The United Nations Office on Drugs and Crime (UNODC) "2017 World Drug Report": https://www.unodc.org/wdr2017/field/Booklet_2_HEALTH.pdf

3. https://www.ncbi.nlm.nih.gov/pmc/articles/PMC3989372/

4. https://qz.com/455977/the-prozac-in-americas-wastewater-is-making-birds-fat-and-shrimp-reckless/

5. https://phys.org/news/2018-08-antidepressants-animal-behaviour-technology.html

O-BOOKS

SPIRITUALITY

O is a symbol of the world, of oneness and unity; this eye
represents knowledge and insight. We publish titles on general
spirituality and living a spiritual life. We aim to inform and
help you on your own journey in this life.
If you have enjoyed this book, why not tell other readers
by posting a review on your preferred book site?

Recent bestsellers from O-Books are:

Heart of Tantric Sex
Diana Richardson
Revealing Eastern secrets of deep love and intimacy
to Western couples.
Paperback: 978-1-90381-637-0 ebook: 978-1-84694-637-0

Crystal Prescriptions
The A-Z guide to over 1,200 symptoms and their healing crystals
Judy Hall
The first in the popular series of eight books, this handy little
guide is packed as tight as a pill bottle with crystal remedies
for ailments.
Paperback: 978-1-90504-740-6 ebook: 978-1-84694-629-5

Shine On
David Ditchfield and J S Jones
What if the aftereffects of a near-death experience were
undeniable? What if a person could suddenly produce
high-quality paintings of the afterlife, or if they
acquired the ability to compose classical symphonies?
Meet: David Ditchfield.
Paperback: 978-1-78904-365-5 ebook: 978-1-78904-366-2

The Way of Reiki
The Inner Teachings of Mikao Usui
Frans Stiene
The roadmap for deepening your understanding of the
system of Reiki and rediscovering your
True Self.
Paperback: 978-1-78535-665-0 ebook: 978-1-78535-744-2

You Are Not Your Thoughts
Frances Trussell
The journey to a mindful way of being, for those who want
to truly know the power of mindfulness.
Paperback: 978-1-78535-816-6 ebook: 978-1-78535-817-3

The Mysteries of the Twelfth Astrological House
Fallen Angels
Carmen Turner-Schott, MSW, LISW
Everyone wants to know more about the most misunderstood
house in astrology — the twelfth astrological house.
Paperback: 978-1-78099-343-0 ebook: 978-1-78099-344-7

WhatsApps from Heaven
Louise Hamlin
An account of a bereavement and the extraordinary
signs — including WhatsApps — that a retired
law lecturer received from her deceased husband.
Paperback: 978-1-78904-947-3 ebook: 978-1-78904-948-0

The Holistic Guide to Your Health
& Wellbeing Today
Oliver Rolfe
A holistic guide to improving your complete health,
both inside and out.
Paperback: 978-1-78535-392-5 ebook: 978-1-78535-393-2

Cool Sex
Diana Richardson and Wendy Doeleman
For deeply satisfying sex, the real secret is to reduce the heat,
to cool down. Discover the empowerment and fulfilment
of sex with loving mindfulness.
Paperback: 978-1-78904-351-8 ebook: 978-1-78904-352-5

Creating Real Happiness A to Z
Stephani Grace
Creating Real Happiness A to Z will help you understand
the truth that you are not your ego
(conditioned self).
Paperback: 978-1-78904-951-0 ebook: 978-1-78904-952-7

A Colourful Dose of Optimism
Jules Standish
It's time for us to look on the bright side, by boosting
our mood and lifting our spirit, both in our interiors,
as well as in our closet.
Paperback: 978-1-78904-927-5 ebook: 978-1-78904-928-2

Readers of ebooks can buy or view any of these bestsellers by
clicking on the live link in the title. Most titles are published
in paperback and as an ebook. Paperbacks are available in
traditional bookshops. Both print and ebook formats are
available online.

Find more titles and sign up to our readers' newsletter at
www.o-books.com

Follow O-Books on Facebook at **O-Books**

For video content, author interviews and more, please subscribe to our YouTube channel:

O-BOOKS Presents

Follow us on social media for book news, promotions and more:

Facebook: O-Books

Instagram: @o_books_mbs

X: @obooks

Tik Tok: @ObooksMBS

www.o-books.com